Environmental Management on North America's Borders

NUMBER FOURTEEN:
Environmental History Series
Dan L. Flores, General Editor

ENVIRONMENTAL MANAGEMENT ON NORTH AMERICA'S BORDERS

EDITED BY RICHARD KIY
AND JOHN D. WIRTH

Texas A&M University Press
College Station

Copyright © 1998 by The North American Institute
Manufactured in the United States of America
All rights reserved
First edition

The paper used in this book meets the minimum requirements
of the American National Standard for Permanence
of Paper for Printed Library Materials, z39.48-1984.
Binding materials have been chosen for durability.

Library of Congress Cataloging-in-Publication Data

Environmental management on North America's borders / edited by
Richard Kiy and John D. Wirth.
  p. cm. — (Environmental history series ; no. 14)
  Includes bibliographical references and index.
  ISBN 0-89096-832-2. — ISBN 0-89096-843-8 (pbk.)
  1. Environmental management—North America. I. Kiy, Richard.
II. Wirth, John D. III. Series.
GE320.N7E58   1998
363.7'056'097—dc21   98-8291
                                                               CIP

*In memory of Timothy Atkeson
(1927–94)*
*Friend and mentor at the United States Environmental
Protection Agency. He was committed to the greening
of the North American Free Trade Agreement.*

Contents

List of Illustrations   IX
Preface   XI
List of Abbreviations   XIII

    Introduction   3
      Richard Kiy and John D. Wirth
1. The Dynamics of Transboundary Environmental Agreements in North America: Discussion of Preliminary Findings   32
      Roberto A. Sánchez-Rodríguez, Konrad Von Moltke, Stephen Mumme, John Kirton, and Donald Munton

PART 1. The North: Canada–United States Case Studies

2. The British Columbia–Washington Environmental Cooperation Council: An Evolving Model of Canada-U.S. Interjurisdictional Cooperation   53
      Jamie Alley
3. Achieving Progress in the Great Lakes Basin Ecosystem and the Georgia Basin–Puget Sound Bioregion   72
      R. Anthony Hodge and Paul R. West
4. Great Whale: From Conflict to Joint Planning of Québec's Energy Policy   108
      David Cliche, with Lucie Dumas

PART 2. The South: Mexico–United States Case Studies
5. Managing Air Quality in the Paso del Norte Region 125
   Peter Emerson, Carlos F. Angulo,
   Christine L. Shaver, and Carlos A. Rincón
   Appendix A:
     Annex VI to the La Paz Agreement 153
   Appendix B:
     Agreement Establishing Joint Advisory 158
     Committee
6. Crossborder Environmental Management and the
   Informal Sector: Ciudad Juárez Brickmakers' Project 164
   Allen Blackman and Geoffrey J. Bannister
7. Carbón I/II: An Unresolved Binational Challenge 189
   Mary Kelly
8. The Maquiladoras and the Environment 208
   Ann C. Pizzorusso
9. Milagro Beanfield War Revisited: Low-level
   Hazardous Waste Sites in Del Rio, Dryden,
   and Spofford, Texas 221
   Alfredo Gutiérrez, Jr.
10. The Handling of Hazardous Industrial Waste on the
    U.S.-Mexico Border: A Case Study of Titisa 234
    Sergio Estrada Orihuela and Richard Kiy

Conclusion 247
   Richard Kiy and John D. Wirth

North American Environmental Chronology 275
Selected Bibliography 287
List of Contributors 289
Index 293

ILLUSTRATIONS

MAPS
Five Major Canada-U.S. Surface Water Areas *Page* 73
Great Lakes Basin 76
Georgia Basin–Puget Sound Bioregion 80
Philips Electronics: Maquiladora Plants 211

CHARTS
Framework of Governance for Great Lakes
   Basin Ecosystem *Page* 88
Framework of Governance for Georgia Basin–
   Puget Sound Bioregion 89

Preface

In keeping with the mission of the North American Institute, this book is the product of a true trinational collaboration. In September 1995, several of the authors met at the Sol y Sombra Conference Center in Santa Fe, New Mexico, to discuss the architecture of the book and to agree on a common set of issues and questions. Now, more than a year later, the edited manuscript is complete. Putting together such a book is always a challenge, and in this trinational project the editors discovered that, indeed, distinct national styles of procrastination exist. Now that all is said and done, however, the final product is a testament to everyone's desire to bring this unique venture to fruition and to establish a broadly North American style of collaboration.

Various chapters were read and commented upon by colleagues. In particular, we wish to thank Sanford Gaines, Jonathan Plaut, and Tim Douglas for their astute and helpful readings. For a grant supporting the book project, we are most grateful to the David and Lucille Packard Foundation. At Sol y Sombra, the Center for the Study of Community provided the ideal venue for our meetings.

ABBREVIATIONS

ACES	Alert Citizens for Environmental Safety, Hudspeth County, Texas
B.C.	Province of British Columbia, Canada
BECC	Border Environment Cooperation Commission
BTU	British Thermal Unit
BZTF	Border Zone Task Force, Philips Electronics North America Corporation
CARE	Citizens Against Radioactive Environment, Kinney County, Texas
CCME	Canadian Council of Ministers of the Environment
CDP	Committee for Popular Defense (Comité de Defensa Popular), Mexico
CEC	North American Commission for Environmental Cooperation, NAFTA
CEMDA	Mexican Center for Environmental Law
CFCs	Chlorofluorocarbons
CFE	Comisión Federal de Electricidad, Mexico
ChemWaste	Chemical Waste Management Corporation, based in Illinois
CNA	Mexican National Water Commission
CONCAMIN	Confederation of Mexican Chambers of Industry
COREMA	Regional Environmental Commission
CTM	Federation of Mexican Workers (Confederación de Trabajadores Mexicanos)

CUSEC	Canadian-U.S. Environmental Council
D.F.	Federal District (Distrito Federal), Mexico
DUGHEST	Deep Underground High Explosive Test
ECC	British Columbia–Washington Environmental Cooperation Council
EDF	Environmental Defense Fund
EIS	Environmental Impact Statement
ENGO	Environmental Nongovernmental Organization
EPA	U.S. Environmental Protection Agency
FEMAP	Mexican Federation of Private Health Associations and Community Development
FNOC	National Federation of Citizens' Organizations (Frente Nacional de Organizaciones Ciudadanas)
FOE	Friends of the Earth
FWS	Fish and Wildlife Service, U.S. Department of Agriculture
GAN	Grupo Acero del Norte
GDP	Gross Domestic Product
GINGOs	Government Induced Nongovernmental Organizations
GLFC	Great Lakes Fisheries Commission
GLU	Great Lakes United
GLWQA	Great Lakes Water Quality Agreement
GONGOs	Government Organized Nongovernmental Organizations
GPO	U.S. Government Printing Office, Washington, D.C.
IAQMB	International Air Quality Management Basin
IAQMD	International Air Quality Management District
IBEP	Integrated Environmental Plan for the U.S.-Mexican Border
IBWC	International Boundary and Water Commission
ICC	Inuit Circumpolar Council
IFC	International Finance Corporation
IJC	International Joint Commission on the U.S.-Canada Border
INE	National Institute for Ecology (Instituto Nacional de Ecología), Government of Mexico
IRP	Integrated Resource Planning
ISO	International Standards Organization
JBNQA	James Bay and Northern Québec Agreement
JPAC	Joint Public Advisory Committee, Commission for Environmental Cooperation of NAFTA

LANL	Los Alamos National Laboratories
LULU	Locally Undesirable Land Use
MOU	Memorandum of Understanding
NAAQS	National Ambient Air Quality Standards
NADBank	North American Development Bank
NAFEC	North American Fund for Environmental Cooperation
NAFTA	North American Free Trade Agreement
NAMI	North American Institute, a trinational public affairs nongovernmental organization based in Santa Fe, New Mexico
NAPP	National Packaging Protocol
NDA	U.S. Nuclear Defense Agency
NEPA	National Environmental Policy Act, U.S.
NGO	Nongovernmental Organization
NIMBY	Not in My Back Yard
NOMs	Official Mexican Standards (Normas Oficiales Mexicanas)
NPCA	National Parks and Conservation Association
NPS	National Park Service, U.S. Department of the Interior
NRDC	National Resources Defense Council
NSPS	New Source Performance Standards
NWF	National Wildlife Foundation
NYPA	New York Power Authority
OECD	Organization for Economic Cooperation and Development
ORCA	Organización Ribereña Contra la Contaminación del Lago de Pátzcuaro
PAHO	Pan American Health Organization
PAHs	Polyaromatic Hydrocarbons
PAN	National Action Party (Partido Acción Nacional)
PCBs	Polychlorinated Biphanols
PEM	Ecologist Party of Mexico (Partido Ecologisto de México)
PNWER	Pacific Northwest Economic Region
PPS	People for Puget Sound
PRI	Institutional Revolutionary Party (Partido Revolucionario Institucional), Mexico
PROFEPA	Federal Attorney General's Office for Environmental Protection, Mexico
PT	Worker's Party (Partido del Trabajo), Mexico
PVOs	Private Voluntary Organizations

RACT	Reasonably Available Control Technology
RAPs	Remediation Action Plans, IJC
RFF	Resources for the Future, Washington, D.C.
SCAT	Sanderson Citizens Against Toxins, Sanderson, Texas
SECOFI	Federal Ministry of Commerce and Industry, Mexico
SEDESOL	Secretariat of Social Development, Environmental Regulation and Enforcement, Mexico
SEDUE	Secretariat of Ecology, Urban Development, and Housing, Mexico
SEMARNAP	Secretariat of Environment, Natural Resources, and Fisheries, Mexico
SIP	State Implementation Plan
TCPS	Texas Center for Policy Studies, Austin
TNRCC	Texas Natural Resource Conservation Commission
TITISA	Tratamientos Industriales de Tijuana S.A., a subsidiary of Chemical Waste Management (ChemWaste)
TRIP	Transboundary Resources Inventory Program
UNAM	National Autonomous University of Mexico (Universidad Nacional Autónoma de México)
USCIB	United States Council for International Business
UTEP	University of Texas at El Paso

Environmental Management on North America's Borders

INTRODUCTION

RICHARD KIY AND JOHN D. WIRTH

The increasing integration of the countries of North America is causing important changes in economic, social, and environmental relationships among Canada, the United States, and Mexico. The increased volume of trade that NAFTA both facilitates and promotes is focusing attention on complex environmental challenges, including transboundary impacts. While a number of these transboundary environmental issues antedate NAFTA, which became operative only in 1994, the advent of this historic trade accord has provided the occasion for considering a range of discrete phenomena as an integrated whole. In fact, what has happened is nothing less than a paradigm shift—a new way of looking at things from a continental, or regional, perspective.

From this regional standpoint, it becomes clear that more is going on than simply a parade of discrete and disconnected issues occurring at specific times in different places on two borders which until very recently few would have said had much in common. We see the transboundary cases as building blocks for a new continental community.[1] Furthermore, we maintain that the habits, lessons, and experiences developed locally are both cumulative and consequential for North America as a whole. The phrase, "Think globally, act locally," is cited frequently, but another thought is equally important: "Act locally, but think regionally and learn how to act globally." In sum, the case studies can be read both as examples of how to manage the local, subnational, and transborder consequences of economic growth, and as harbingers of larger things to come.

It is clear that NAFTA has strengthened cooperation among the three North American nations on environment and natural resource issues. Bilateral methods and institutions for protecting the environment, some of them dating back to the beginning of this century, are being reshaped by powerful regionalizing trends, while NAFTA has inspired the governments to create new trinational institutions, including labor and environmental commissions. Indeed, the pace of economic integration is such that a range of issues now can be solved *only* trinationally and cooperatively. These issues include the long-range transport of airborne pollutants, problems of water management and contamination, and the accelerating effects of highway and rail traffic on land use and air quality along the new north-south transportation corridors.

While the spirit and reality of continental cooperation are strong, another dynamic is at play. Expanding trade and investment flows across borders, including the growth of urban centers along the borders, inevitably will produce environmental stresses that easily can spawn policy conflicts, not only among federal governments but also among state, provincial, and municipal jurisdictions. The ways in which conflict and cooperation are shaping the actions of citizens and governments managing crossborder environmental issues constitute the thematic thread unifying this collection of case studies.

Managing complex environmental issues is inherently conflictual. Where trade and environment effects intersect, difficult issues of implementation and jurisdiction are being raised. This linkage of trade and environment issues remains controversial, and in 1997 a study by NAFTA's Commission for Environmental Cooperation (CEC) to monitor the environmental effects of free trade was delayed for months because officials at the Mexican Ministry of Trade were concerned about the possible effects that publishing these findings might have on trade.[2] Yet to be developed is a policy mechanism to solve environmental disputes cooperatively before they spill over into trade conflicts. Meanwhile, the requisite majority of Canadian provinces, which have jurisdiction over the environment, still has not ratified the North American Agreement for Environmental Cooperation, the side agreement to the main NAFTA treaty.

This complex dynamic of conflict and cooperation shows up most vividly along the continent's internal borders. On the southern border of the U.S., the depletion of natural resources shared with Mexico—water, especially—has led to seemingly intractable problems and disputes. And on the northern border with Canada, problems of air and water quality persist despite a massive and partially successful effort, starting in the 1970s, to clean up the Great Lakes. The adverse environmental effects of rapid growth in the maquila industries (in-bond assembly plants[3]) on the U.S.-Mexican border were a spur to producing the side agreement to the main NAFTA treaty. Water quality issues are being addressed by the states and by two bilateral institutions created under NAFTA, the Border Enviroment Cooperation Commission (BECC) and the North American Development Bank (NADBank). Much Remains to be done, however.

Current transborder conflicts over enviromental protection and natural resources run the gamut of issues, in both subject and scope:

1. Long-range transport of pollutants by air and sea is continental in scope but often manifests itself in local or regional conflicts. Examples are acid rain carried into Canada from the U.S. and to New England from Canada, and pollution from Mexico that affects the ecosystems of the Gulf of Mexico and the Great Lakes.
2. Discharges from the massive Carbón coal-fired power plants at Piedras Negras may be degrading visibility at Big Bend National Park. This circumstance has created a dispute involving the U.S. Environmental Protection Agency (EPA), the U.S. Department of the Interior, the State of Texas, and the Government of Mexico, as discussed in Mary Kelly's chapter. In the U.S., recent deregulation of electrical transmission lines gives a price advantage to coal-fired plants in the Midwest over plants in the East, fueled by oil and natural gas. Ontario is likely to suffer from increased emissions of sulfur dioxide (SO_2), sulfur oxide (SO_x), and nitrous oxide (NO_x).
3. Proposed, and now canceled, expansion plans for Québec power plants pitted the Cree Nation of Canada, environmental nongovernmental organizations (ENGOs), and the state governments of New England and New York against the Québec provincial power authority. This classic case of conflict and cooperation is profiled by Québec Minister David Cliche.
4. Sewage flows from Tijuana into the San Diego metroplex led to the 1983 Border Cooperation Agreement between the U.S. and Mexico and are still being dealt with at local, state, and federal levels. The "La Paz Agreement," as it is called, provides a useful framework for consultation on water, air, and the transport of hazardous materials. In fact, water issues figure prominently in the EPA's new Border XXI Program to develop and implement remediation projects with Mexico.[4]
5. Victoria, the capital of British Columbia, Canada, still discharges only partially treated sewage into the Straits of Juan de Fuca, giving rise to complaints from Puget Sound communities to the south. Victoria's policy is at variance with a massive binational effort to curtail municipal discharges into the Great Lakes, as discussed by Tony Hodge and Paul West in this volume.
6. The proposed siting of low-level radioactive waste facilities along the Texas border has spawned alliances between local governments on both

sides of the frontier, and also involves ENGOs and Mexican and American authorities, both federal and state. Conflicts of the type referred to as NIMBY ("not in my back yard") are prevalent, and the dispute over the siting of a low-level nuclear waste landfill at Sierra Blanca, Texas, is still unresolved, as the chapter on Tratamientos Industriales de Tijuana (TITISA) by Sergio Estrada and Richard Kiy and the chapter by Mayor Alfredo Gutiérrez of Del Rio, Texas, discuss.

7. With U.S. production of chlorofluorocarbons (CFCs) now phased out, a growing and highly profitable trade in smuggled Freon coolant for use in older model cars is moving from Mexico, where production is still legal under the Montréal Protocol, to the U.S. and Canada, where it is illegal. In addition, poorly monitored and underreported shipments of industrial hazardous waste continue to find their way into Mexico, as the chapter by Estrada and Kiy points out.

8. Extending the U.S. Gulf intracoastal waterway to Tampico would threaten to degrade the Laguna Madre, a major marine breeding ground and bird sanctuary which is part of the Rio Grande estuary. A binational alliance, including the Texas Center for Policy Studies and the Grupo de los Cien, a Mexican environmental group, currently is fighting the waterway in Mexican courts. Pollution of the San Diego Bight is a local manifestation of the continental deterioration of marine habitat. Other estuaries are becoming stressed, including the Georgia Bay–Puget Sound region, which is profiled in this book.

9. Fisheries disputes, seemingly a constant of North American relations, range from difficult and complex negotiations over the harvesting of salmon and tuna in the Pacific, to the impact of Mexican shrimping practices on the Kemp's Ridley and Loggerhead sea turtles in the Gulf of Mexico, to the seizure of American boats by Canadian and Mexican authorities. Spanish fishermen sailing under the Mexican flag have been seized by Canadian patrol boats in the North Atlantic. The recently resolved tuna-dolphin dispute between the U.S. and Mexico has important implications for the new World Trade Organization, where the linkage between trade and environment issues is under discussion. Because of the salmon dispute, relations between British Columbia and the State of Washington are experiencing stress. Implicated is the British Columbia–Washington Environmental Cooperation Council, which is profiled here by Jamie Alley.

To be sure, with globalization, transboundary environmental conflicts no longer are limited to the borders. Already, international ENGOs have forced British Columbia to modify its timber harvesting practices. Exportadora de Sal (a joint venture of the Mexican government and Mitsubishi), which is already the world's largest producer of industrial salt, is under pressure from ENGOs in Mexico and the U.S. to halt its plans to build a new saltworks at San Ignacio Lagoon. The lagoon, located on the Pacific coast of the Mexican state of Baja California Sur, is a prime breeding ground of the California grey whale, a shared resource of Canada, the U.S., and Mexico. Similarly, plans by the U.S. Department of Interior and the State of Alaska to sell lease rights for oil exploration in the Beaufort Sea are generating opposition from environmental groups and indigenous populations in both the U.S. and Canada who are concerned about the potential impact on binational feeding areas of the bowhead whale, the beluga whale, the Porcupine Caribou, and migratory birds.

The increase of economic activity on the borders, it may be argued, is a great spur to political and social liberalization, raising public awareness and enhancing public desire to improve the quality of life. However, given the increase of trade in North America and the emergence of broad public concerns over the environment, conflicts such as these are inevitable, and the need to resolve them is growing. Above all, four factors give direction and momentum to transborder environmental management across North America.

First, the public in all three countries shares a perception of environmental stress shaped by such sensational incidents as the "death" (eutrophication) of Lake Erie in the late 1960s, the flows of Tijuana sewage onto San Diego beaches, and the crises of polluted airsheds in Los Angeles and Mexico City in the 1980s.

Second, the public's perception of environmental stress has fueled the emergence of the environment as a political issue. Demands for action have accelerated, starting thirty years ago in Canada and the U.S., with some success in the Great Lakes region, and later in Mexico, following the government's mismanagement of Mexico City earthquake relief in 1985.

Third, intergovernmental institutions—such as the International Joint Commission (IJC) on the northern border, and the advent of NAFTA and its side agreements, which focused attention on the southern border—have provided space for a host of transborder initiatives: by governments at all levels, by business, and by private voluntary organizations including ENGOs, whose transborder networks have proliferated rapidly.

Finally, transboundary problems once viewed as local and discrete more and more are seen, by both governments and the public, as connected to the global ecosystem, and as important—indeed, inescapable—aspects of their economic and political relations.

Undoubtedly, the passage of NAFTA raised public awareness concerning environmental and natural resource problems along common borders. Areas once peripheral to large nations now are seen as emerging regions in their own right. That they are regionally interconnected is less understood. Bilateral, unconnected, and fragmentary activities on the part of the three federal governments and other authorities are still the rule—as Roberto Sánchez and his colleagues point out in their chapter. Even so, this is changing, because the three NAFTA partners have similar environmental problems although they do not have the same levels of resources to address them. They do share the need for new institutions to manage these problems, and they have an aroused and interconnected public.

This leads us to another common theme: the experience of environmentalism is helping to create the social base for a North American community. At the public consultations on North American issues organized by the Joint Public Advisory Committee (JPAC) of the CEC, for example, participants seek out like-minded people in the process of commenting and exchanging ideas. This process has been advanced by the North American Fund for Environmental Cooperation (NAFEC), which, since its creation by the CEC in 1995, has encouraged community groups and ENGOs to submit joint proposals for binational and even trinational projects. To date, NAFEC has made grants of almost four million (Canadian) dollars for collaborative projects on the environment and sustainable development. Public participation is also encouraged on the CECNet, a recently inaugurated Internet discussion site which follows the lessons learned and success of BECCNet, a program of the Border Environment Cooperation Commission (BECC).

BECC is a new binational agency which relies heavily upon public comment in certifying infrastructure projects which are then sent to the North American Development Bank (NADBank) for funding approval. In promoting dialogue, these new institutions provide means for the funding applicants, commentators, and other participants to create community. Public advisory committees have been formed. For example, the U.S. Good Neighbor Environmental Board, representing many stakeholders, and the Region 1 Council of the Mexican Advisory Council for Sustainable Development have begun to meet jointly to discuss border issues. The EPA's U.S.-Mexico Border XXI Program has been crafted with the aid of extensive public comment. The same can be said for the binational working groups of the Western Governors Association and for the extensive consultations on the other border which take place under the umbrella of Great Lakes United.

The conclusion we draw is that, under NAFTA, these trade partners, with their quite different federal systems, now share similar problems of environmental management, institutions which intersect, and an aroused and interconnected social base which is forming a North American community. Furthermore, socioeconomic change on the borders is running far ahead of the capacity of governments at all levels to meet increasing demands for services. Given this situation, informal relationships and practical problem solving often fill the gap. In the process, practices and techniques are developed which then can be applied elsewhere on North America's borders. This puts a premium on networking, varied approaches, and flexibility. The upshot is that this emerging community may resemble, as much as anything, a honeycomb of mutually dependent and cooperative relationships more informal than formal, but gaining in density and structure.

What follows is a discussion of common values, some thoughts about specific conditions on the borders, a brief history of trinational perspectives, and a concluding section on policy implications, including the role of civic and business organizations.

THE ECOLOGICAL PERSPECTIVE

By no means do Canadians, Americans, and Mexicans all share the same interests or even desire the same outcomes; the effects of different histories and different resource capacities see to that. What seems to have happened, however, is that the "ecological perspective," which "has become widely shared among better informed, better educated, and more affluent Canadians and Americans," now has been embraced by their Mexican counterparts as well. According to Lynton Keith Caldwell, an astute analyst of this politics, "In its purest form, the ecological perspective clashes with conventional political and economic values across a broad range of specifics involving attitudes toward economic growth, legal rights (particularly in the ownership and use of property), national sovereignty, and the necessities of national security." Moreover, he continues, "The ecological perspective on life and the world for North Americans tends to be continental in a global context."[5]

Since Caldwell wrote these lines in the mid-1980s, the ecological perspective has been embraced by a much broader public, and transborder collaboration among like-minded citizens has burgeoned, now on the U.S.-Mexican border as well. In their chapter, Hodge and West show how such a value structure can nurture a strong sense of community and appropriate related institutions on the Canada-U.S. border. Although the battles with conventional views on growth are far from over, in the public's eye economy and environment are yoked together. What seems to be emerging is the "continental-ecological frame of reference" which Caldwell hoped would take environmental management beyond discrete, often reactive problem solving and into the deeper waters of coordinated, systemic actions to solve regional problems in a global context.[6]

One of the first fruits of this continental framework is NAFTA's Commission for Environmental Cooperation (CEC), based in Montréal.[7] The CEC's program to monitor and measure the sources of continental air pollution eventually may lead to a trinational air quality management authority.[8] The chapter by Peter Emerson and his colleagues on the Paso del Norte Air Management Region profiles what could become one of the local, crossborder building blocks. In February, 1997, the long-established International Joint Commission on the U.S.-Canada Border (IJC) held its first meeting ever on the southern border, so as to compare notes on air management with colleagues from Ciudad Juárez and El Paso. The San Diego–Tijuana metropolitan region now is considering a similar air management plan.

A program for the comparative analysis of North American estuaries, with attention to managing the land-based sources of pollution for marine environments, is in the works. Programs for a North American Toxic Release Inventory and for the sound management of hazardous chemicals have been approved by the three governments. And, as of this writing, a trinational agreement to establish a transboundary environmental impact assessment is pending approval by the U.S. State Department.

On the borders, it often seems that states and provinces are moving ahead of the federal governments in their search for solutions to shared environmental problems. The Gulf of Maine Commission is a good example. The State of Texas has taken the lead in establishing the Transboundary Resources Inventory Program (TRIP), the first systematic geographical information survey of the entire southern border. Having become accustomed to acting cooperatively on a range of environmental health and infrastructure problems, citizens in border cities express impatience with the failure of state and federal agencies to keep pace. Civic and business groups in the San Diego–Tijuana have created a de facto metropolitan area, but solutions to such problems as goods clearance and air pollution caused by trucks and automobiles waiting at the border are in the hands of federal officials who are more concerned with drug and immigration issues than with solving local problems. The same can be said of other border cities. In fact, social and economic integration has outrun the political institutions, creating a serious lag in the capacity of government to deal with transborder management issues. However, it is on the borders, and most particularly in the many examples of environmental collaboration among a range of new actors and jurisdictions, that the pieces of an emerging North American environmental tapestry can be perceived most clearly.

THE BORDERS

The northern border separates the people of two wealthy countries with similar socioeconomic profiles, for whom the environment has been a major concern for at least a generation. A long history of popular concern with transborder issues began at least as early as 1885, when the State of New York, in cooperation with the Province of Ontario, established the Niagara Reservation to protect Niagara Falls and its surrounding binational ecosystem from destruction by hydro projects and rampant commercialization. In both countries, population, which is stable in the eastern and central reaches of the border, is growing rapidly in the West. There a type of metropolitan, or Cascadian, corridor is being created from Vancouver to Eugene. Since 1980, cross-border networks have proliferated, especially in the Great Lakes region under the umbrella of the IJC. Here on the northern border, as on the southern, the concept of ecomanagement has developed beyond reactive policies and been institutionalized; but for the moment, due to budgetary constraints, true sustainability may not be achieved.

The southern border is shared by neighbors who generally are not well off financially. In fact, some of the poorest communities in the U.S. are strung out along this border. Yet, for Mexicans migrating northward from Central Mexico in search of jobs, the border, with its burgeoning *maquiladoras* (assembly plants), is seen as the land of growth and economic opportunity. Since 1965, 3,839 such plants, with 980,000 employees, have been established in Mexico, 70 percent of them on the border. Now well into a population boom—ten million people and growing by 4 percent a year—this border features a string of cities interspersed with long stretches of thinly populated lands and small rural communities. Tijuana, across from San Diego, California, and Ciudad Juárez, across from El Paso, Texas, are two of the fastest growing cities in North America.

Once considered economically and politically peripheral, border communities no longer can be taken for granted. However, income levels remain low, and substandard housing is common. Infrastructure development has not kept pace with the rapid growth of the border areas. Many residents lack adequate housing, potable water, wastewater treatment, and solid waste disposal systems; and substandard living conditions result in much higher than normal illness rates.[9]

Bilateral agreements specific to this border date from the 1889 Convention. Signed between the two federal governments to deal with problems arising from shifts in the course of the Rio Grande and Colorado rivers, the convention resulted in the resolution of the long-running Chamizal dispute. For many years, concern for the shared environment was limited to a narrow band of issues pertaining to sovereignty and water allocation. Previously a policy backwater, this "other" border came into its own in the early 1990s because of NAFTA. Crossborder networks and institutions are emerging here, but some twenty years behind their well-developed counterparts on the northern border.

While environmental management in the North deals fundamentally with the consequences of industrialization—as much with problems created by prior growth as with the probable consequences of future growth—environmental management in the South deals with the ongoing effects of recent, accelerating industrialization and the anticipated consequences of future growth. If the one community reflects the concerns of a postindustrial society, the other reflects the concerns and interests of a young, rapidly industrializing population emerging from poverty. Are these two quite separate binational relationships, with little in common? The studies in this volume suggest not. Countervailing forces of integration are at work, thanks to the development of a continental market, the new NAFTA institutions, and the recent development of a common storehouse of ideas and techniques available for transborder problem solving. On the two borders, networking by subnational governmental units, business interests, and ENGOs shows more similarities than differences. Shifting demographics on both borders, coupled with growth brought about through expanded trade and commerce, contributes to the new reality. And in all three countries, a shift in power away from national capitals to subnational units and transborder regions is under way. Moreover, the development of an ecological perspective in key sectors of the public in all three countries provides a shared consciousness, a social base, and a North American constituency for action on a continental scale.

A HISTORICAL PERSPECTIVE

The earliest efforts in North America to plan a shared environment go back more than a century, but until the 1970s the institutions to resolve transboundary issues and disputes were narrow in scope and limited in large part to questions of water allocation, conservation of natural resources, and bilateral agreements. (See chronology at the back of this book.) In the years before 1970—before ENGOs, environmentalism, an aroused public, and federal agencies charged specifically with protecting the environment, such as Environment

Canada, EPA, and SEDUE (later SEMARNAP, the Mexican environmental ministry)—relatively little was accomplished, although the elements of an agenda appeared early on. This agenda, centered on the conservation of natural resources, emerged in an era of internationalism and globalizing forces from the late 1880s to the Balkan wars of 1912. Then, after a long hiatus, a greatly expanded agenda reemerged in the 1980s, as we entered another era of liberal internationalism.

In 1894, the U.S. convened the first Irrigation Congress in Denver, Colorado, with representatives of both Canada and Mexico.[10] At this congress, the idea was broached that a multilateral organization should be established to manage North America's shared natural resources, including water. At a following meeting, held the next year in Albuquerque, New Mexico, the question of the equitable distribution of transboundary water resources along the Rio Grande came up. In an effort to press the U.S. on similar resource allocation issues along the U.S.-Canadian border, the Canadian delegate joined with his Mexican colleagues in a resolution asking the U.S. to support the "appointment of an international commission to act in conjunction with the authorities of Mexico and Canada in adjudicating the conflicting rights which have arisen, or may hereafter arise, on streams of an international character." When pressed by the British ambassador (who handled Canada's interests in Washington), the U.S. State Department, cautious about losing influence over such matters if the U.S. were an equal partner with its two neighbors, backed away.[11]

In the absence of U.S. support for an international authority to resolve transboundary water issues, Canada focused on a bilateral approach, seeking support for an institution similar to the International Boundary Commission which had been established in 1889 between the U.S. and Mexico to resolve boundary disputes along the Rio Grande. In 1905, the International Waterways Commission was duly created to advise the two governments on water and boundary issues. From this grew the International Joint Commission (IJC), established in 1909, which finally, in the 1970s, came into its own as an important umbrella organization and coordinating agency for the massive effort to clean up and manage the Great Lakes. Meanwhile, the Mexican boundary commission became the International Boundary and Water Commission (IBWC) in 1944, maintaining its narrow focus on boundary and water questions until the 1970s, when its scope expanded to embrace water quality and sewage treatment issues. Then the pace of institutional change on the border accelerated greatly under NAFTA.[12]

While official Washington preferred to deal one on one with Canada and Mexico when it came to water matters, President Theodore

Roosevelt had a much wider vision of continental cooperation. In 1908, Roosevelt called the governors of U.S. states to Washington for a White House conference on conservation, which for the first time established as a conservation goal the protection of human health from the dangers of pollution. In the waning months of his presidency, Roosevelt promoted the North American Conservation Conference, which brought representatives of Canada, Mexico, and the U.S. to Washington in February, 1909. That conservation was a problem that transcended the boundaries of single nations was the major conclusion of this congress, and another conceptual breakthrough. A subsequent conservation congress of world leaders was to have been held at the Hague, Netherlands; but Roosevelt's successor, William Howard Taft, concerned with domestic fiscal issues and balancing the budget, canceled it. Thus was lost an early opportunity to establish a framework for an international regime to protect the global commons. Likewise, the North American Conference itself had little lasting impact, as Taft turned away from continental issues, the Porfirio Díaz regime was overthrown in Mexico's great Revolution of 1910, and Premier Wilfred Laurier lost the pivotal 1911 election to Canadian nationalists over the issue of free trade with the U.S.[13] An era of liberal internationalism was ending, and its spirit would not reappear until the 1980s.

Later, the three countries found ways to promote cooperation and minimize transboundary disputes, mostly on the basis of bilateral agreements, except for the U.S-Canada-Mexico Migratory Bird Treaty of 1936. (For a checklist of agreements and key dates, see the chronology at the back of this book.) Important milestones include the U.S.-Mexico Water Treaty in 1944, governing the allocation of surface water along the Rio Grande, the Colorado, and the Tijuana rivers; and the Border Environmental Cooperation Agreement (also known as the La Paz Agreement), signed in 1983 by Presidents Miguel de la Madrid and Ronald Reagan. La Paz contained the first modern transborder air agreement: Annex IV governed SO_2 emissions from copper smelters on the Arizona-Sonora border. On the other border, Canada and the U.S. signed the Great Lakes Water Quality Agreement in 1972, sixty years after the first study of transboundary water pollution. This was followed in 1991 by the U.S.-Canada Air Quality Agreement, known as the Acid Rain Accord.

The La Paz Agreement originated with President Ronald Reagan's desire to fulfill an earlier campaign promise to the citizens of California to secure Mexico's support for cleaning up the sewage-ridden Tijuana River, which flows north from Mexico to the cities of Imperial Beach and San Diego. For his part, President Miguel de la Madrid was eager to join Reagan in signing a cooperative agreement, so as to gain influence in Washington in the aftermath of the 1982 debt crisis and during the darkest days of the Contra Wars in Central America. Signed on August 14, 1983, in La Paz, Baja California Sur, the agreement contained an annex specifically addressing the Tijuana River problem. In the next few years, the agreement's scope was broadened to encompass border binational emergency response (Annex II), transboundary movement of hazardous waste (Annex III), smelters (Annex IV), and urban air pollution (Annex V).[14]

La Paz is a classic example of how the existence of a framework can open opportunities for a broad range of loosely related actions. This agreement institutionalized regular consultations among government officials at the ministerial level and created a framework for expert groups to report on highly contentious and politically charged environmental issues. Thus La Paz undoubtedly facilitated trust and openness among officials of the two governments to a marked degree in the 1980s.

Regional agreements in other spheres, such as the European Community's embrace of environmental regulations, the United Nations' environmental programs for less developed countries, and the U.N.-sponsored Earth Summit in Rio de Janeiro in 1992, with its resultant framework conventions, helped to improve acceptance of linkages between trade and the environment. However, because of Mexico's lingering economic troubles, as well as a lack of resources and will on both sides of the border effectively to address common environmental problems, with the exception of smelter emissions, relatively little, or relatively slow, progress has been made on the specific commitments set forth in the annexes to the La Paz agreement.

Some bilateral initiatives were kept alive by purely personal actions. When officials reached an impasse as to who ultimately was responsible for cleaning up some illegally dumped U.S. waste on the Mexican side, Tim Atkeson, then U.S. national coordinator under La Paz and the EPA's assistant administrator for international activities, used his own American Express card to cover the cost of proper disposal.

It is well worth asking why La Paz, as an available framework, was not used to address the incipient problem of Mexican coal-fired power plants on the Texas border before they became the major issue that they are today, as Mary Kelly relates. Why, indeed, were state officials, ENGOs, and other elements in civil society so slow to react to this pollution threat, in contrast to their highly effective role ten years earlier in seeing that emissions from the Arizona and Sonora copper smelters were controlled?[15] If approved by the three countries, the transboundary environmental impact assessment would help to head off future industrial pollution problems.

With the 1991 U.S.-Canada Acid Rain Agreement, official Washington finally recognized that the transborder transport of stack gasses is a problem, and in large part a problem created in the U.S.—a breakthrough after years of stalling by the Reagan administration. This accord commits industry and both governments to achieve emission targets and abatement goals. Due to intense public concern, this was Canada's premier bilateral issue during the 1980s. However, Canadian officials were reluctant to confront the Americans with more than one major issue at a time. Passage of the Canada-U.S. Free Trade Agreement in 1988 cleared the way for them to tackle acid rain. Meanwhile, after ten years of debate, Congress's passage of the 1990 Clean Air Act amendments committed the U.S. to stringent domestic emission quotas, but only after long and complex negotiations among environmental groups, eastern and western coal producers, and electric utilities in the Midwest. Then the agreement with Canada was signed, ending a major irritant in the relationship but by no means ending the problem. Large sums still must be committed, but the emissions trading program appears to be an efficient and economical way to allocate the costs and benefits of abatement. Ironically, Canada, the junior partner when it comes to emissions, has done relatively less than the U.S. since 1991, which puts it in the posture of becoming a free rider. Ontario is concerned about the probable effects of deregulated transmission systems on the U.S. side, which is only the most recent aspect of a systemic problem.

THE CHANGING POLICY LANDSCAPE

As of this writing, it is clear that cooperation on air issues is progressing. What is really new is the will to view the problem continentally. How, for example, can toxics in Great Lakes waters be addressed without dealing with the long-range air transport of certain pollutants, including DDT from Mexico and Central America? Big Bend National Park experiences reduced visibility, but the pollution comes from sources as distant as Houston and Monterrey, in addition to the Carbón power plants and dust from unpaved roads. To the extent that Carbón plants add to continental SO_2 loading, Canada is affected as well.

Since the signing of La Paz, environmental conditions on the southern border have worsened. There has been a population surge, as a growing number of U.S. and other multinational companies have begun locating their manufacturing and assembly operations along the border, to take advantage of Mexico's lower wage rates, good transportation, and proximity to large consumer markets. Within six years after La Paz, the number of maquilas almost doubled, to over 1,700. Despite expectations that they would be phased out under NAFTA, the number of maquilas has continued to grow, especially after Mexico's 1995 monetary devaluation reduced wage rates. Several of the newest plants are not merely assembly plants for parts manufactured in the U.S. and Japan, but rather are full state-of-the-art manufacturing plants.

In November, 1990, with environmental conditions worsening, the U.S. and Mexican presidents met in Monterrey to discuss what could be done to improve conditions on the border. To be sure, the main purpose of the November, 1990, presidential summit between George Bush and Carlos Salinas de Gotari was to initiate discussions concerning a U.S.-Mexico Free Trade Agreement. What is important to note here is that, from the start, the political need to address border environmental concerns was not lost on the two presidents. Trade and border environmental issues in fact were not linked officially until the spring of 1991—only when that linkage was forced by outside pressure. But they were discussed, if not associated, from the start.[16] When Canada joined the discussions in February, 1991, making the negotiation trilateral and leading to the NAFTA agreement signed in 1993, these well-publicized border environmental problems provided strong arguments for political leaders, ENGOs, and progressive industry environmental advocates to insist upon the explicit linkage of trade and environment in the agreement. Also, a review of U.S.-Mexico environmental issues was undertaken by the U.S. trade representative, with the cooperation of EPA. Released on February 29, 1992, concurrently with the Border Plan enunciated by President George Bush, the environmental review was intended to respond to criticism that the NAFTA negotiations were not addressing likely environmental impacts. At the same time, the review identified key issues that needed to be addressed by the governments. In this way NAFTA evolved into the first trade treaty to take account of environmental protection; in short, it became "the first Green trade treaty."

Passage of NAFTA by the U.S. Congress depended upon adding green provisions to the treaty while insuring adequate enforcement of each country's environmental laws—all in response to public concerns and the demands of environmental groups. The approach which emerged from negotiations over the environmental side agreement is geared more to cooperative agenda setting and capacity building than to enforcement, in order to accommodate the specific concerns of Mexico and Canada as well as other interest groups. The CEC itself was established under the authority of the North American Agreement on Environmental Cooperation, signed at the White House by the three environmental ministers on September 19, 1993, less than two months before congressional passage of NAFTA.[17]

Responding to growing political pressure in the U.S. over concerns that liberalized trade with Mexico could result in further degradation of the environment, Bush and Salinas committed their environmental agencies to developing a joint action plan for the border. However, the administration's reluctance to undertake a full-scale environmental impact assessment while the trade agreement was under negotiation led to a lawsuit brought by Public Citizen, the consumer action group, and joined by the Sierra Club. Although these ENGOs lost their court battle in the summer of 1993, just months before the historic NAFTA vote, their actions put pressure on both the Bush and the Clinton administrations. However, without the pressure by groups supporting NAFTA to include significant environmental components in the NAFTA accord, it is doubtful that NAFTA would have turned out as "green" as it did.[18]

Both supporters and opponents of NAFTA, including many labor and environmental groups, continue, as they did during the treaty debate, to use the border as their litmus test for measuring the likely effects of trade flows under NAFTA.[19] According to a recent government study, much needed work still is not being done, and business has not followed through on its responsibility to address important problems on the border.[20] As of this writing, it is fair to say that the Clinton administration has missed opportunities to alleviate environmental problems on the borders. In fact, the attitude in Ottawa, Washington, and Mexico City seems to be that NAFTA has been done. This disengagement from NAFTA issues masks disagreements within the three federal governments, some agencies of which (such as Mexico's Trade Ministry) never have accepted the trade-environment linkage in NAFTA. Meanwhile, the requisite number of Canadian provinces still have not ratified NAFTA, and footdragging by the U.S. State Department has slowed the momentum of institutional development on the U.S-Mexico border. Lack of follow-through and advocacy has slowed the implementation of NAFTA-generated programs, opening the way for the same groups who opposed NAFTA in 1993 to demonize Mexico and adopt what amounts to a nationalist position on environmental matters. The "Green Group" of ENGOs that supported NAFTA with the side agreement was inactive four years later when the environment again was discussed along with trade. Depressingly enough, the words *cooperation, community,* and *partnership* almost never were used when NAFTA was mentioned in the November, 1997, Congressional debates over "fast track."[21] Playing constituent politics with the trade issue, members of the U.S. House of Representatives seemed oblivious to realities on the border.

The first Border Plan was soon sidelined in the 1992 presidential election, and funds committed by the Mexican Congress in support of the plan were not matched by the U.S. Congress. Consequently, the border communities reaped few benefits. Furthermore, the lack of extensive public hearings on both sides of the border before the plan's release was criticized strongly by nongovernmental organizations (NGOs). In another oversight, the plan did not pay sufficient attention to health issues. These problems are being addressed in the current Border XXI program which EPA developed with considerable public input, all of which reflected the heightened interest and engagement of border citizens in border issues today.[22]

For all its problems, the first border plan was a necessary first step in drawing up a comprehensive action plan for a volatile region that just then was coming under public scrutiny. Above all, the first plan signified the border's high profile in North American affairs, something totally new for this once marginalized and neglected region. NAFTA also provided a large institutional umbrella for an agreement creating two new bilateral agencies: BECC, which helps communities develop plans for, and then certifies, border-area environmental infrastructure projects; and NADBank, which provides the loan capital for BECC-certified projects. After a slow start, these binational institutions were working well in 1997.

Also important was the mid-1980s emergence of the environment as an issue in Mexican politics, accompanied by the rapid growth of ENGO networks on the border and propelled by the very strong, pent-up demands by Mexican citizens as they increasingly joined the global economy to participate in a more open and democratic civil society. With its open public hearings, the BECC itself, since its inception in 1995, has become an important vehicle of expression and community building.

MODES OF COOPERATION: ROLE OF CIVIC AND BUSINESS GROUPS

Government and industry increasingly must give due consideration to the environmental laws of neighboring nations, especially when enterprises or projects have transboundary environmental impacts. A notable early example is the famous Trail Smelter dispute on the British Columbia–Washington State border which established in international law the principles that a nation has the responsibility to insure that activities under its jurisdiction do not damage the environment of another country, and that, when they do, the polluter pays.[23] Subsequently, a growing number of cases have arisen in which the environmental laws of a neighbor country or state have directly impacted projects which traditionally would have been subject only to domestic laws and regulations.

One case in point is a June, 1993, ruling of the California Superior Court, which imposed a $2.5 million judgment against RSR Industries of Dallas and Quemetco, its Los Angeles-area subsidiary, for illegally dumping lead slag at a Tijuana recycling and smelting plant operated by Alco Pacific. Alco also was fined $2.5 million for having shipped and disposed of lead batteries illegally there. While the court held that Quemetco had not caused direct environmental damage in California, it ruled that the company had violated the state's hazardous waste shipping and handling laws by illegally disposing of the waste in another country.[24] Significantly, Tijuana residents had filed a class-action lawsuit against Quemetco in the State of Texas, which provides standing for foreign nationals to join suits when the state's interests are affected.

In 1994, a different kind of lawsuit was filed in Texas by Mexican and U.S. nationals in their efforts to head off plans for the Spofford hazardous waste site. Mexican nationals also participated in the Texas permitting hearings in a few other site applications.[25] The important role of transborder coalitions and their court actions in disputes over siting is vividly delineated by Mayor Antonio Gutiérrez in his chapter in this book.

In another sign of the changing policy landscape, state and provincial authorities, in association with private groups, are accumulating experience while addressing transnational issues of common concern. Jamie Alley, a civil servant in British Columbia, Canada, provides an excellent case study in his chapter on the origins and development of the British Columbia–Washington Environmental Cooperation Council. In addition, the management of marine estuaries has been pioneered under the Gulf of Maine Agreement, a provincial-state initiative which was signed in December, 1989, by the premiers of New Brunswick and Nova Scotia and the governors of Maine, New Hampshire, and Massachusetts. The agreement and accompanying action plan are not limited to marine waters, but adopt an ecosystem approach covering coastal areas and watersheds in the region and encompassing land-based pollution as well.[26]

Great Lakes United is a binational citizens' coalition of some two hundred organizations which is engaged in a broad range of programs aimed at conserving and protecting the ecosystem, including monitoring water quality, fisheries, and natural resources. For its part, the 1972 Great Lakes Water Quality Agreement fostered "the largest regularly occurring ... environmental forum in North America" until it was canceled recently for budgetary reasons (see Sánchez). At the community level, the potential for new partnerships among many different stakeholders is being harnessed in the recently established Alliance of Border Communities, a coalition of towns and cities in British Columbia's Fraser Valley and Washington State's Whatcom and Skagit counties. Mayors, business leaders, environmentalists, representatives of each of three community colleges, and arts leaders serve on the binational board.[27]

If North America is becoming a storehouse of ideas, practices, and techniques for transborder environmental management, no more compelling sign of the emerging North American community exists than in the astonishing growth of ENGOs. Starting some thirty years ago on the northern border and developing rapidly since 1990 on the southern border, these groups have established binational and trinational alliances on issues of common concern, such as the preservation of migratory species and the transboundary movement and incineration of hazardous waste. According to David Ronfeldt, senior social scientist at the RAND Corporation, and Catherine Thorup, senior policy advisor with the U.S. Agency for International Development, "All across North America, the realm of civil society has been long dotted by isolated interest groups that normally acted independently, and with little communication, consultation, and coordination with each other. But this has been changing rapidly. Today, the spread of crossborder networks and coalitions of NGOs promises to restructure the landscape of civil society all across North America."[28]

This is a new kind of Information-Age activism, based on non-hierarchical, flexible organizations among NGOs "concerned with modern and postmodern issues such as the environment, human rights, immigration, indigenous people, cyberspace, etc."[29] ENGOs and other private voluntary organizations (PVOs) are children of the information revolution, using faxes and the Internet to foster connectedness—exchanging information, concentrating their forces, seeking out like-minded groups and individuals. Just as NAFTA and the new institutions have expanded the policy space available for dealing with transborder issues, so ENGOs use information to expand their roles as change agents interacting with governments and communities, and as shapers of public opinion on environmental issues.

Thanks to the Internet, small groups and individuals can make a difference by focusing the power of information and ideas at the right time and place. A case in point was the successful effort to save the NADBank appropriation in late 1995. Mark Spalding, a border attorney, was following the on-line record of a Senate subcommittee when he noticed that $54 million of paid-in and callable capital for the fledgling bank had just been zeroed out. Spalding responded by mobilizing a binational constituency through BECCNet, the list server of the University of Arizona's Udall Center, which tracks the BECC, and the cut was restored.[30]

ENGOs are not a specifically American phenomenon. This model of civic action transfers well to Canada and Mexico, despite marked differences in political culture.[31] Associational energies of this sort can flourish on the southern border, and this book's two case studies from the Juárez–El Paso area provide good examples. ENGOs, then, cannot be dismissed as culture-specific or elitist, and the emergence in recent years of populist, grassroots activism is important, as is discussed in our conclusion to this book. ENGOs have emerged as major contributors to the development of a North American community.

The ENGO phenomenon has many facets, from "brick throwing," monitoring compliance with environmental laws, and entering partnerships and coalitions, to brokering the complex mix of different interests involved in a given project.[32] With its tradition of building consensus through roundtables, Canada has much experience in handling environmental conflict. In their eagerness to participate and to create the underpinnings of a democratic society, Mexican groups are forcing changes based on greater public participation and the glimmerings of community right-to-know legislation. Amid the current wave of reports concerning loss of "social capital," declining ability to process information, and diminishing will to associate, among contemporary Americans at least, this style of activism, once identified for what it is, offers a refreshing and powerful alternative model. Far from "bowling alone" (the evocative—and withering—metaphor of political scientist Robert Putnam), these North Americans hear the trumpet of Keith Caldwell's "ecological perspective" and are actively attuned to the role of government and public purposes.[33] They are seeking each other out; they are protecting the environment together; and in the process they are helping to create the sinews of a North American community. The CEC's Joint Public Advisory Committee saw this interaction at its 1996 environmental hearings in Montréal, San Diego, and Toronto. Trinational commentary on BECCNet provides additional strong evidence for it. Are these manifestations mere pockets of engagement, or are they harbingers of broader changes in civil society? Could it be that the Mexicans, through interacting with Americans and Canadians in ENGO networks, will spur their post-industrial colleagues to new forms of civic commitment?

To be sure, Mexican environmental activism is a relatively new phenomenon. It was kindled in 1985 in the wake of such disasters as the explosion of PEMEX storage tanks in San Juan Ixhuátepec, in the State of Mexico; and the Mexico City earthquake, when the deaths of thousands highlighted the lack of environmental and safety standards, as well as federal government ineptitude in contrast to the effective relief efforts of Mexican and foreign PVOs. The Grupo de los Cien (Group of 100), an elite group of leading artists, writers, and scientists, "came into being on March 1, 1985, in protest against pollution and environmental negligence in the Valley of Mexico and the complete lack of official and industrial attention to these problems." Acting on behalf of all citizens in the city and country they were defending, the group's first concrete victory took place in June, when it successfully opposed expansion of the Mexico City airport into Lake Texcoco, a local issue. Later it took on projects farther afield, such as supporting Tijuana citizens opposing TITISA, protecting the Lacondona rain forest, and bringing the Cozumel pier case to the CEC.[34]

Environmental activism on the border did not surface until the early 1990s, when the prospect of NAFTA's passage spotlighted the border's environmental shortcomings. In an important example of grassroots activism, the Tijuana Housewives Association allied with foreign ENGOs in 1992 to prevent establishment of TITISA, a fifty-million-dollar incinerator project. The Projecto Fronterizo de Educación Ambiental had its roots in the binational effort to control smelter pollution on both sides of the border in the late 1980s, before moving on to address hazardous waste and water issues. An example of community activism on the U.S. side was the struggle by the citizens of Del Rio, Texas, to prevent the siting of toxic waste dumps on their aquifer. This campaign was bolstered by the active participation of Mexican officials and concerned citizens. And in 1992, the Group of 100 and the Alert Citizens for Environmental Safety (ACES) formed a binational coalition to oppose the siting of a low-level nuclear waste dump in Sierra Blanca, Texas. The Group of 100 is mustering opposition to the Laguna Madre canal project, in alliance with the Texas Center for Policy Studies.

Problems well known to border residents, such as the environmental mismanagement of certain maquila plants and the high incidence of infants born with neurological brain defects in the Matamoros-Brownsville area, had received little attention until the national news media and ENGOs focused on these issues and their possible connections to NAFTA. Thereafter, federal agencies from both countries began targeting large sums for border projects. Heightened public concern in the three countries over the environmental effects of NAFTA led the U.S. and Mexican governments to promise more than six billion dollars to clean up the border. Some of the six billion dollars includes World Bank financing to Mexico (which was provided), and state revolving loan funds are available on the U.S. side. However, only four hundred million dollars of dedicated funding has been committed for infrastructure projects to date; and the NADBank, with two billion dollars of paid-in and callable capital is under pressure to devise more flexible loan packages for impoverished municipalities unaccustomed to meeting the terms of long-term loans at market rates.

The irony for Mexico is that less than 6 percent of its entire population lives on the border, and it can be argued that many other environmental problems deserve priority. The opportunity for Mexican ENGOs and their transborder allies lies in the overarching structure provided by NAFTA and its new institutions, coupled with the availability of public and foundation funding. Also contributing to this opportunity is the totally new experience (for Mexican localities) of public hearings, of which the BECC provides a prominent example. The peso's collapse in 1995 greatly reduced the support available to ENGOs from Mexican funding sources and institutions, a setback which was especially hard on local activists and their fledgling organizations along the border. Mexican government restrictions on the press, including censorship and subtle coercion, still pose problems to ENGOs seeking greater access to information. However, the Internet is changing this somewhat, and Mexican officialdom is becoming more open.

The hope in Mexico is that concerns about the border will engage Americans' sustained attention and understanding of the serious environmental problems in other parts of Mexico, such as air quality and a looming water crisis in the Valley of Mexico. To this extent, engagement on environmental issues plays into the entire range of issues in the complex, maturing relationship between the two neighbors. Canada is involved peripherally but has a large stake in providing environmental services to Mexico while Canadian ENGOs are actively networking with their Mexican and U.S. counterparts.

For their part, Canadian ENGOs, who had been more comfortable using government funds than their colleagues in the other two countries, are having to adjust to the effects of budget downsizing and cutbacks at Environment Canada. Just as Mexican environmental groups have turned away from government paternalism—they used to be called GONGOs, for government-organized nongovernmental organizations—so Canadian groups without public subsidy no longer run the risk of being GINGOs, or government-induced nongovernmental organizations.

The corporate community is changing, too, moving from a reliance on government to remedy the political and social consequences of pollution to a more proactive stance.[35] Fostering environmental management is not limited to Canadian and American companies, as Ann Pizzorusso points out. She chronicles the efforts of a European-based multinational to bring its maquilas into compliance with companywide global standards by promoting better environmental management in border plants. One tool is training to meet the ISO 14000 voluntary environmental standards. This involves not only training managers but also getting employees at all levels to buy into the new practices. The proposition is that good environmental management means better management overall, which translates into more competitive operations.

Based in Europe, the International Standards Organization (ISO) sets a variety of performance, quality, and environmental standards, with the 9000 program widely accepted, and the more demanding 14000 program now coming in, especially at the multinational and large company level. Meeting the ISO standards is a way to achieve voluntary compliance with environmental regulations set by governments. In North America, meeting the ISO 14000 standard may one day also become a de facto green labeling for industry as consumers seek out environmentally responsible producers. In Canada, the U.S., and Mexico, ISO 9000 quality standards are widely adopted by multinational and other large companies, especially those exporting abroad, but companies have been slower than their European counterparts to adopt the banner of ISO 14000. Some export-oriented Mexican companies like CEMEX, the cement manufacturer, which is growing overseas, see economic benefits from adopting ISO 14000. Caught in a credit squeeze, many smaller Mexican companies worry about meeting the extra costs of compliance with ISO 14000.

Mexican government officials, who are accustomed to regulating by command and control techniques and have been pressing companies to conduct their own environmental performance audits, have been skeptical of ISO 14000. Some ENGOs see the ISO movement as a possible way to weaken regulations by setting private standards for various industries. But internal mechanisms to achieve voluntary compliance with environmental laws constitute a powerful new tool, brought to the border in 1996 in workshops endorsed by the CEC and sponsored by industrial organizations of the three countries. Just as business is learning to operate in a public space transformed by environmentalism and trends running deep in the civil society of North America, so also the ENGOs and government officials must grant space for the growth and development of environmental culture and advocacy within the corporations. Having said this, it is clear that business responds most often when other players in the public and private spheres bring pressure.[36]

It is here, at the intersecting roles of corporations, government, and voluntary groups, where some of the most interesting aspects of conflict and cooperation in dealing with the continental environment will be worked out. Public-private partnerships are the order of the day, but if the erstwhile partners turn hostile toward business, it may be because business interests have failed adequately to communicate their intentions and practices to other stakeholders in their respective communities. Or it may be because business is reverting to earlier views on the environment. Recent attempts by conservative politicians and many businesses to secure regulatory relief in the U.S., Canada, and Mexico have triggered a firestorm of opposition from the public, expressing its core belief in environmental protection.[37]

Having focused on the changing social and institutional landscape across North America and on the growing influence of environmental activism, we now invite readers to explore the case studies for specific examples. The authors of several chapters were active in the events they describe. In a sense, they are writing history as they made it and saw it being made. If this approach carries with it the risk of a certain lack of balance and objectivity, as one of the readers of this manuscript pointed out, a strength lies in having the perspective of participant observers and in understanding how they saw the unfolding of complex events in the heat of controversy. Other points of view are cited in the footnotes. The cases themselves offer premier examples of the complex and conflicting interests on North America's borders. Have the editors, however, chosen the right cases? To be sure, others might have chosen differently, but for our purposes the cases work well. Besides, most of this is new material, replete with "breaking" news. To stimulate more research on border environmental topics, the editors have included in their conclusion a group of suggestions for future research.

We hope this book will encourage a broader understanding of the dynamics that are shaping decisions on transboundary environmental issues. The lessons learned from the case studies are relevant not only in North America, but also in our neighboring nations in Central and South America, and especially in Mercosur, the South American trading bloc where transboundary issues also are crucial. And as governments begin to deal with the economic and social consequences of fulfilling their commitments to freer trade and to global agreements such as Agenda 21 and the International Framework Convention on the Reduction of Greenhouse Gas Emissions, they will recognize the importance of regional solutions to environmental problems. North America provides a storehouse of available experiences—ideas, examples, and practices for others to examine and then to adapt to their own specific circumstances and conditions, should they choose to do so.

NOTES

1. For a discussion of the North American community, see Robert L. Earle and John D. Wirth, eds., *Identities in North America; The Search for Community* (Stanford, Calif.: Stanford Univ. Press, 1995). See also John D. Wirth, "Advancing the North American Community," *American Review of Canadian Studies* 26, no. 2 (Summer, 1996): 261–73.
2. Anthony DePalma, "NAFTA Environmental Lags May Delay Free Trade Expansion," *New York Times,* May 21, 1997.
3. Mexican in-bond assembly plants receive components from the U.S. or other foreign countries free of duty, which are then assembled into finished goods for re-export to the host country, again free of duty except for the value of the labor added.
4. EPA, "U.S.-Mexico Border XXI Program, Draft Framework Document" (Washington, D.C.: GPO, June, 1996).
5. Lynton K. Caldwell, "Binational Responsibilities for a Shared Environment," in *Canada and the U.S.: Enduring Friendship, Persistent Stress,* ed. Charles F. Duran and John H. Sigler (New York: Prentice-Hall, 1985), 218. For one appraisal of the political implications of Caldwell's thought, see Mebs Kanji, "North American Environmentalism and Political Integration," *American Review of Canadian Studies* 26, no. 2 (Summer, 1996): 183–84.
6. Caldwell, "Binational Responsibilities," 221–24. On the existence of broad-based public attitudes toward the environment, see Kanji, "North American Environmentalism," 198–99.
7. See Pierre Marc Johnson and André Beaulieu, *The Environment and NAFTA: Understanding and Implementing the New Continental Law* (New York: Island Press, 1996); Stephen P. Mumme and Pamela Duncan, "The CEC and the U.S.-Mexican Border Environment," *Journal of Environment and Development* 5, no. 2 (June, 1996): 197–215; Joseph F. DiMento and Pamela M. Doughman, "The NAFTA Environmental Side Agreement Implemented," mimeographed paper, Global Peace and Conflict Studies, Univ. of California, Irvine, Apr., 1997; and Sanford E. Gaines, "The CEC Is Working, But Don't Bring Other Countries In," *Environmental Law Forum* 14, no. 5 (Sept.–Oct., 1997): 41–43.
8. CEC, Annual Program and Budget, 1996. In the first stages of the air program, the three governments will work toward establishing data compatibility and the same technologies for monitoring air toxics, including

joint placement and calibration of specialized equipment at locations mutually agreed upon. CEC, *Continental Pollutant Pathways: An Agenda for Cooperation to Address Long-Range Transport of Air Pollution in North America* (Montréal: CEC, 1997).

9. The conditions are described in a global context in Helen Ingram, Nancy K. Laney, and David M. Gillian, *Divided Waters: Bridging the U.S.-Mexican Border* (Tucson: Univ. of Arizona Press, 1995), ch. 1.

10. Chirakairkan Joseph Chacko, *International Commission Between the U.S. and the Dominion of Canada,* Studies in History, Economics and Public Law no. 358 (New York: Columbia Univ. Press, 1932), 71.

11. U.S. Senate, Document no. 253, 54th Cong., 1st sess. (May 11, 1896), 3; Chacko, *International Commission,* 73.

12. The 1909 Boundary Waters Treaty between the U.S. and Great Britain, requiring that water not be polluted and health and property not be damaged on either side of the border, was a first in international law. For a brief discussion of the IJC and its transformation, consult R. A. (Tony) Hodge and Paul R. West, "Achieving Progress in the Great Lakes Basin Ecosystem and the Georgia Basin–Puget Sound Bioregion," this volume. On the IBWC, see Stephen P. Mumme, "State Influence in Foreign Policymaking: Water-Related Environmental Disputes Along the U.S.-Mexico Border," *Western Political Quarterly* 38, no. 4 (Dec., 1985): 620–40. After a long period of tension, relations between the U.S. and Mexico became less strained in the mid-1980s. By that time, the IBWC was virtually the only bilateral agency at the federal level. Today it cooperates closely on water infrastructure issues with the BECC and the NADBank, two new NAFTA institutions on the southern border.

13. Paul Russell Cutright, *Theodore Roosevelt: The Making of a Conservationist* (Urbana: Univ. of Illinois Press, 1985), 231; and Philip Shabecoff, *A New Name for Peace: International Environmentalism, Sustainable Development, and Democracy* (Hanover, N.H.: Univ. Press of New England, 1996), 20–21. The early transborder conservationism of Theodore Roosevelt foundered on such mundane factors as sovereignty issues, budget cutting, and domestic concerns.

14. Key staff members included Cliff Metzner from the U.S. State Department, and Alberto Székeley from Secretariat of Foreign Relations (SRE). The Environmental Defense Fund played a leading role in Annex IV. As Peter Emerson and his colleagues relate in this volume, the Paso del Norte air initiative could have fit into the La Paz framework under a specially created Annex VI, but the Mexican authorities preferred a lower-profile venue.

15. John D. Wirth, *Smelter Smoke in North America: Transborder Air Pollution and the Regional Commons,* in progress.

16. President Bush committed himself to linkage on May 1, 1991. Letter, Sanford Gaines to John D. Wirth, Oct. 26, 1996.

17. Pierre M. Johnson and Beaulieu, *Environment and NAFTA.* The CEC had its roots in the Bush administration. In September, 1992, EPA Administrator William P. Reilly hosted a meeting of environmental ministers at Blair House, at which he and his counterparts Luis Donaldo Colosio and

Jean Charest agreed to establish the North American Commission on the Environment, which, after a year of intense negotiations, became the current CEC.

18. As Sanford Gaines, formerly of the office of the U.S. Trade Representative, relates, the environmental impact statement lawsuit had no effect on negotiations and policies. The administration saw it as a potential obstacle to approval under the "fast-track" process, but the pressure applied by Public Citizen and the Sierra Club was effective only on food safety issues. The side agreement was affected by pressure from potential NAFTA supporters, including EDF, NRDC, NWF, and the Audubon Society. The Sierra Club and Public Citizen, because predisposed to oppose all of NAFTA, remained outside the CEC negotiating process.

19. E.g., see Public Citizen (a Nader group), *NAFTA's Broken Promises: The Border Betrayed: U.S.-Mexican Border Environment and Health Decline in NAFTA's First Two Years* (Washington, D.C.: Public Citizen Publications, in association with the Red Mexicana de Acción Frente al Libre Comercio, Jan., 1996); executive summary is available on the Internet at www.citizen.org/pctrade/broken_pr.html.

20. U.S. House of Representatives, Committee on Commerce, "International Environment: Environmental Infrastructure Needs Along the U.S.-Mexico Border Region Remain Unmet," Report to the Committee's Ranking Minority Member (Washington, D.C.: General Accounting Office, 1996).

21. Under fast-track authorization, the Congress must vote up or down on the entire trade agreement as negotiated by the president. This prevents special interests from riddling a previously negotiated agreement with favors, exemptions, wavers, and/or from blocking parts of it. Conferred by Congress, fast-track authority expired in January, 1995, and for political reasons has not been renewed, to the joy of protectionists and the dismay of free traders, who wanted Chile in NAFTA.

22. EPA and SEDUE, *Integrated Environmental Border Plan for the U.S.-Mexico Border, First Stage, 1992–94* (Washington, D.C.: GPO, 1992) was critiqued by the plan's Public Advisory Committee. See "State of the U.S.-Mexico Border Environment," prepared for EPA by the Udall Center, Univ. of Arizona, Sept., 1993. The next border plan relied upon extensive consultations: EPA, *Border XXI, Domestic Meeting Summary Report, U.S.-Mexico Border Communities* (Washington, D.C.: GPO, Apr., 1996); and Richard Opper, "Border Progress: Illegal Arms, Bad Strawberries, and a Faltering Foundation," *NAMI News* (Fall, 1997): 1–3.

23. John D. Wirth, "The Trail Smelter Dispute: Canadians and Americans Confront Transboundary Pollution, 1927–1941," *Environmental History* 1, no. 2 (Apr., 1996): 34–51.

24. The result of a binational investigation, these fines and imprisonment of Alco Pacific's chief executive marked the largest criminal settlement in an environmental matter in California's history. See Diane Lindquist, "Executive Gets Prison for Abandoning Toxic Lead in Baja," *San Diego Union Tribune*, Dec. 15, 1993, A-1, and Sandra Dibble, "Cleanup is Set for Toxic Site in Tijuana," *San Diego Union Tribune*, June 27, 1996, B-1. Almost

three years after the ruling, remediation still has not begun, and what is proposed is an asphalt capping operation to cover the tons of lead waste onsite, "a frustratingly ineffective remedy" which, according to attorney Richard Opper, has been delayed by years of bureaucratic footdragging on the Mexican side. Opper to Richard Kiy, Oct. 15, 1996.

25. Texas Center for Policy Studies, "Proposals for Border Waste Dumps Spur Unprecedented Binational Interest," feature article in *Ambiente Fronterizo* (Texas Center for Policy Studies, Austin) (Apr., 1994): 5–8.

26. Aldo Chircop, David VanderZwaag, and Peter Mushkat, "The Gulf of Maine Agreement and Action Plan," *Marine Policy* 19, no. 4 (1995): 317–33.

27. Memo, Joy Monjure, Columbia Pacific Foundation, to John Wirth, Sept. 13, 1996. The first chair is Tim Douglas, former mayor of Bellingham, Wash., and now director of community, trade, and economic development for Washington State.

28. David Ronfeldt and Catherine Thorup, "NGOs, Civil Society Networks, and the Future of North America," in *Trans-Border Citizens: Networks and New Institutions in North America*, ed. Rodney Dobell and Michael Neufeld (Lantzville, B.C.: Oolichan Books, 1994), 21–39.

29. John Arquilla and David Ronfeldt, *The Advent of Netwar* (Santa Monica, Calif.: RAND Corporation, 1996), 71.

30. See Mark J. Spalding, "The NADBank Appropriations: Battle Is Won; Does the War Go On?" *BECC/COCEF Perspectivas* 1, no. 3 (Jan., 1996): 1, 11. To be sure, BECC and NADBank staff also used conventional lobbying techniques to restore the appropriation; these were equally, if not more, important.

31. On ENGOs in Canada, see Doug Macdonald, *The Politics of Pollution* (Toronto: McClelland and Stewart, 1991); on Mexico, consult E. Kürzinger et al., *Política ambiental en México: El papel de las organizaciones no gubernamentales* (Mexico City: Fundación Friedrich Ebert, 1991). See also Jolle Demmers and Barbara Hogenboom, "Popular Organization and Party Dominance: The Political Role of Environmental NGOs in Mexico," doctoral thesis, Dept. of International Relations, Univ. of Amsterdam, Netherlands, Nov., 1992. An inventory is found in Centro Mexicano para la Filantropía, *Directorio de instituciones ambientalistas en la República Mexicana* (Mazatlán, Mexico: Centro Mexicano para la Filantropía, 1996). For an astute assessment of U.S. groups, see Christopher J. Bosso, "After the Movement: Environmental Activism in the 1990s," ch. 2 in *Environmental Policy in the 1990s*, ed. Norman J. Vig and Michael E. Kraft (Washington, D.C: Congressional Quarterly Press, 1994). For a useful chronology of groups and issues, consult Benjamin Kline, *First Along the River: A Brief History of the U.S. Environmental Movement* (San Francisco: Arcada Books, 1997). And for emerging environmental movements in Latin America, consult Victor M. Toledo, "Latinoamérica: Crisis de civilización y ecología política," *Gazeta Ecológica* 38 (Spring, 1996): 13–22.

32. See Rebecca R. Bannister, John D. Wirth, and Rod Dobell, eds., "The Role of Environmental Non-Governmental Organizations (ENGOs) in

North America," report from a workshop held by the North American Institute, Santa Fe, N.M., Feb., 1995.

33. Robert D. Putnam, "Bowling Alone: America's Declining Social Capital," *Journal of Democracy* 6, no. 1 (Jan., 1995): 65–78, is a provocative and highly influential article by a political scientist. The *Washington Post*, in association with the Kaiser Foundation and the Harvard Univ. Survey Project, published an extremely pessimistic report on American civic attitudes, in a series of articles running in the *Post* from Jan. 28 to Feb. 4, 1996. See also "Why Don't Americans Trust Their Government?" (Menlo Park, Calif.: Kaiser Family Foundation, 1996).

34. Information on the Grupo de los Cien Internacional, its official name, was provided in E-mail, Betty Aridjis to John Wirth, Jan. 7, 1997. The director is Homero Aridjis, a writer who recently was elected president of PEN International.

35. See "The Fun of Being a Multinational," *Economist* 340, no. 7975 (July 20, 1996): 51. Also, Denis Smith, "Business and the Environment: Toward a Paradigm Shift?," in *Business and the Environment: Implications of the New Environmentalism*, ed. Denis Smith (London: Paul Chapman Publishing, 1993), 1–11.

36. To be sure, this is not without risk for the larger ENGOs, with their national and regional agendas. They may find themselves bypassed by corporate environmental officers who have become adept at dealing one-on-one with local groups, thus localizing (and containing) issues. See Michael McCloskey, "The Skeptic: Collaboration Has Its Limits," *High Country News* 29, no. 9 (May 13, 1996): 7. On the disappointing results of voluntary initiatives to alleviate border environmental problems, consult Richard Opper, "Border Progress: Illegal Arms, Bad Strawberries, and a Faltering Foundation," *NAMI News* (Fall, 1997): 1–3.

37. Leyla Boulton, "Business and the Environment," *Financial Times* (London), June 30, 1995.

1

THE DYNAMICS OF TRANSBOUNDARY ENVIRONMENTAL AGREEMENTS IN NORTH AMERICA

DISCUSSION OF PRELIMINARY FINDINGS

ROBERTO A. SÁNCHEZ-RODRÍGUEZ,
KONRAD VON MOLTKE, STEVEN MUMME,
JOHN KIRTON, AND DON MUNTON

This chapter is suggestive, a starting point for much-needed research using a regional and comparative approach.[1] Here we present some preliminary findings from the first comprehensive study of federal-federal, province-state, state-state, and local-local environmental interactions in North American government relations. The study was carried out for the Commission for Environmental Cooperation (CEC). The CEC's goal for transboundary agreements in North America is to enhance and strengthen efforts to promote cooperation on environmental issues in North America.[2] Specifically, the objectives of our study were the following:

1. To survey and create a database on existing transboundary environmental interactions between federal, state/provincial, and local governments, among the three countries of North America.
2. To analyze these in terms of relevant key characteristics, including participating actors, nature of the agreement, nature of the issues involved, date of origin, etc.
3. To analyze these patterns of interactions in terms of their relationship to the broader context of Canada-U.S.-Mexico relations, and to assess their importance.

The database on transboundary environmental agreements and our review of transboundary environmental activities reveal a rich and complex set of transboundary environmental management activities throughout the North American region. From the evidence, it is clear that national governments, states, provinces, and local governments increasingly are engaged in a spectrum of activities, including pollution control and remediation, environmental health management, conservation (wildlife management and biodiversity protection), and the protection and development of fisheries and agriculture. The range and scope of these activities certainly is impressive and extends to most issue areas that are of significant concern on the contemporary environmental agenda.

Though increasingly embedded in a framework of multilateral and trilateral agreements, North American transboundary environmental management remains predominantly bilateral in nature. This fact is a function of both geography and politics. The North American region is composed of three nations whose boundaries are drawn in a straightforward, fixed, largely undisputed, and binary fashion. This fact is significant in limiting the scope and nature of territorial disputes and those involving natural resources and environmental degradation. The political dimension derives in part from the geopolitical realities of the boundaries; but it also reflects prevailing practices in international law and diplomacy, where bilateral approaches historically have predominated over multilateral forms of cooperative engagement across boundaries.

THE DATABASE

It is assumed that a high proportion of all bilateral and trilateral environmental agreements between the NAFTA parties are included in the current database. No estimate of the degree of completeness is possible since no comprehensive listing of such agreements exists elsewhere. Confidence in the level of completeness is based on the availability of several overlapping compendia and the process of inquiry that underpins the database. However, further agreements likely do exist. This database will seek to incorporate in its next update those agreements not included in this version.

The review of the database suggests that the three nations now have a better institutional mechanism to address questions of a decidedly trinational character across a broad range of issue areas: air quality, hazardous substances, biodiversity—indeed, most of the issues now subsumed under the general rubric of global change, as these affect the North American area. The institutions created during the last few years to address environmental issues also potentially could infuse new resources for environmental management into traditionally underfunded or neglected trinational and binational areas.

Nevertheless, there still are very few trilateral agreements in effect between Canadian provinces and U.S. and Mexican states or between other jurisdictions represented in all three NAFTA countries below the national level. This finding is not surprising, given, first, the lack of forums in which such jurisdictions could articulate the North American environmental dimension of their mission; and, second, the practical obstacles to the development and effective implementation of any such agreement. Moreover, subnational jurisdictions almost by definition are responsible for issues of limited geographical scope, so that the North American environmental dimension of their mission can be expected to be modest, at least insofar as issues not covered by national actions are concerned.

The number of bilateral agreements between North American public authorities other than national agencies is known to be large, and the database therefore is assumed to be incomplete at this time.

The database includes a number of multilateral environmental agreements in which two or more NAFTA parties participate. No attempt has yet been made to develop this aspect of the database systematically, but a number of sources are available that would permit its completion once appropriate criteria for inclusion have been elaborated. A number of issues relating to these multilateral agreements are discussed below.

A review of the database reveals that most issues of environmental management are addressed in one form or another. In policy terms, "the environment" comprises, in fact, a number of interdependent but often independently managed policy areas, including air and water quality, soil protection, waste management, toxins, and ecosystem protection. All of these issues are covered to a certain degree, as are many of the instruments that have been developed to address them: research, education, and environmental assessment. Most of these aspects are addressed by one or more North American bilateral or trilateral agreements.

While the overall level of coverage is comprehensive in terms of issues addressed, it is difficult to identify any area other than the management of transboundary water that can be deemed satisfactorily covered. In most instances, the bilateral or trilateral agreements represent an adjunct to continuing activities of the parties and frequently were established in response to specific circumstances. Most agreements involve some process of meetings, information exchange, and possibly coordination but rarely extend to substantive analysis and priority setting.

A number of issues stand out by virtue of the larger number of agreements that have been concluded to address them—in particular, water pollution, air pollution, and wildlife management in all its aspects.

WATER POLLUTION

In almost all instances, the agreements focus on transboundary water management or on specific international river basins. The result of these agreements is an institutional structure that is fairly well developed and based primarily on the International Joint Commission (IJC) and the International Boundary and Water Commission (IBWC). The former, in turn, has created a significant number of institutions to address issues relating to specific bodies of water which, over the years, have given rise to transboundary disputes on the Canada-U.S. border. The result is a complex structure that has proven itself effective in a number of instances. Evaluations of the water management structure on the U.S.-Mexican border are less clear, since the mandate and composition of the IBWC always have been significantly less strong than those of the IJC. Moreover, the dominant concerns relating to water management differ markedly along each border.

Water policy certainly is the best institutionalized policy category in the border areas of the trinational region and in the region at large. Even so, as seen above, both approaches and dominant concerns vary markedly from border to border. Water relations are well institutionalized on the U.S.-Canada border, particularly with regard to water quality. The IJC historically has had jurisdiction over boundary waters and has been the primary agency of reference for the adjustment of surface water disputes, including those directed at water pollution problems. In the U.S.-Canadian case, plentiful water supplies have directed bilateral concern towards the development of protocols on water levels of the Great Lakes, hydroelectric power projects and their transboundary effects, the water quality of rivers and lakes, and, to a lesser extent, the allocation of waters. Rivers affected include the St. Lawrence, Niagara, St. John's, St. Mary's and Milk, Kootenay, Skagit, and Columbia.

The IJC has primary jurisdiction over boundary waters, narrowly construed, but it does not enjoy exclusive jurisdiction in the area of water resources management and lacks an operational mandate actually to construct and administer projects. Operations and maintenance remain within the purview of domestic agencies at national, state/province, and local levels in both the U.S. and Canada. The IJC's jurisdiction over boundary waters has expanded over the years through various treaty instruments to include the St. Mary's and Milk, which extends its jurisdiction to water allocation in this instance, and the Great Lakes Water Quality Agreement, which extends its jurisdiction to "the Great Lakes ecosystem." Insofar as the IJC is not specifically limited to certain water management functions, its jurisdiction may be construed as potentially broader in scope than that of the IBWC; by contrast, the IBWC's mandate and charge is technically more specific and exclusive under its charter, the 1944 U.S.-Mexico Water Treaty. In general, the IJC thus functions primarily as a diplomatic body for the investigation and management of disputes related to the use of boundary waters and secondarily as an administrative agency.

While criticized for its ad hoc and diplomatic approach to water resources management, from an institutional perspective the IJC long remained the only game in town for transboundary water management. Unlike the situation on the U.S.-Mexico border, on the U.S.-Canada border there is no additional "framework" agreement for addressing environmental issues writ large. Even so, the IJC has been effective in developing a range of institutional responses to particular problems and, in the case of the Great Lakes, has developed a well-articulated structure for comprehensive water quality management in coordination with public agencies and the private sector. The IJC continues to work in both countries with a wide range of domestic agencies with competencies in the water management area, as well as with a wide range of nongovernmental organizations.

It should be noted that much of the IJC's effectiveness hinges upon the willingness of the governments to elect to "refer" issues to the IJC for resolution. In recent years, the two governments have been reluctant to use this avenue, electing instead to utilize independent diplomatic channels.

On the U.S.-Mexico border, the IBWC exercises the leading role in binational policy development and implementation, supplemented now by the BECC-NADBank apparatus. Traditionally, the IBWC has interpreted its mandate to apply strictly to those waters that were indisputably international (having clear transboundary effects) and, in the case of water pollution, to those issues specifically presenting a hazard to human health (e.g. sewage and sanitation problems). Even within this narrow jurisdiction, IBWC coordinates with a wide range of domestic agencies with mandates related to water supply and water quality. In the U.S., these include the Bureau of Reclamation, Army Corps of Engineers, EPA, the Fish and Wildlife Service, and state level agencies; in Mexico, they include SEMARNAP and CNA. In 1983, the IBWC's jurisdiction was supplemented by the U.S.-Mexico Border Environmental Cooperation Agreement (the La Paz Agreement). This agreement provides a generalized framework, under the leadership of the two national environmental agencies, for the development of a binational water policy to work in tandem with the IBWC in those cases in which water issues are decidedly transnational in character. Since 1994, the BECC has coordinated with the IBWC in helping to develop water remediation projects and water management infrastructure. BECC certification is contingent upon projects satisfying environmental and sustainable development criteria.

AIR QUALITY

The weakness of transboundary air quality management is one of the most salient phenomena of international environmental management in North America. The existence of important issues is beyond dispute. These range from acid rain and the long-range transport of toxic pollutants, to the need to manage local air quality in densely populated border regions and the concern for air-pollution impacts on parks. The available structure of agreements and institutions for transboundary air quality management is quite limited, and its effectiveness is questionable. The institutionalization of binational management of air quality is in its infancy, although various problems have been identified and addressed in an ad hoc fashion over the past two decades. Until the late 1970s, air quality management in the trinational area was under the exclusive jurisdiction of domestic agencies. On the U.S.-Mexico border, beginning in 1978 with a landmark Memorandum of Understanding between the EPA and Mexico's Subsecretariat for Environmental Improvement in the Ministry of Health, the two nations have moved to improve coordination of air quality activities along the border. This process was given greater authority and momentum by the La Paz Agreement in 1983. Since then, two annexes to the La Paz Agreement have been signed, Annex IV and Annex V. The Integrated Border Environmental Plan further identified a range of regional air quality problems for study and mitigation, and specified implementation plans for Tijuana–San Diego, Mexicali–Imperial Valley, and Ciudad Juárez–El Paso for the 1992–94 implementation period.

Actual implementation of these air quality agreements remains spotty. Under Annex IV, the "smelter triangle" emissions problems on the Arizona-Sonora border have been reduced substantially, due to closure of the Phelps Dodge smelter at Douglas, Arizona; installation of sulphur dioxide capture technology at Nacozari, Sonora; and adjustment of production schedules at copper smelters at Cananea and Nacozari. Allegations abound, however, that allowances are being exceeded, and continued monitoring and surveillance are necessary to insure compliance.

Annex V implementation has been confined largely to Ciudad Juárez–El Paso. There, under joint authority of EPA and SEMARNAP and with cooperation by local authorities and the Environmental Defense Fund, a three-year project is implementing Annex V's provisions for an in-depth study of the sources and movement of pollutants within the airshed. The project also is studying the possibility of developing an International Air Quality Management District. Implementation of functions outlined by Annex V and the IBWC has lagged in other target regions of the border, however.

Both Canada and the U.S. are signatories to the convention on Long-Range Transport of Air Pollution (1981), an agreement that commits its members to cooperate in the control and reduction of pollutants emitted into the atmosphere. The fact that Mexico is not a party to this agreement creates an obstacle to reaching a comprehensive agreement for the trinational region.

ENVIRONMENTAL HEALTH

Environmental hazards long have been identified as significant health risks in the border area. Thus an informal group, the U.S.-Mexico Border Health Association, was established in 1978 with support from the Pan American Health Organization (PAHO), to address common concerns. The Border Health Association has pursued the development of epidemiological databases, disease registries, and other types of environmental data to include air quality and groundwater data relevant to protecting public health in the border region. Pressed by the Border Health Association, the two countries in 1994 created a new organization, the U.S.-Mexico Border Health Commission. Its mandate is to coordinate responses in the public health sectors at various levels of government and community on both sides of the border to better address health threats arising from such sources. Yet another initiative, the Rio Grande Alliance, is presently being developed as a means of coordinating environmental health activities in the Rio Grande reaches of the border region.

MIGRATORY SPECIES

As with hazardous waste management, the approach to international management of migratory species in North America is primarily multibilateral; that is, cooperation among the three countries is based on a system of bilateral agreements rather than on a coherent trilateral structure or on a broader Western Hemispheric agreement that could provide a framework.

Transboundary cooperation for the conservation and protection of wildlife and biota in the North American region is an important developing field. While significant measures have been taken on bilateral, trilateral, and international bases, sizable gaps still are found in the area of migratory species management, wetlands protection, and other areas.

Conservation and the protection of biodiversity are better developed on the Canada-U.S. border than on the southern boundary, although significant and long-standing measures are found on both borders. In general, the protection of waterfowl and migratory birds has received the greatest attention. Probably the oldest instrument to this end is the Convention for the Protection of Migratory Birds, signed in 1916 among Great Britain, Ireland, and the U.S. An early agreement of its type, the convention does not deal with habitat loss, only with the taking of birds through hunting and other means. A similar agreement in 1936 between the U.S. and Mexico is the Convention Between the United States and the United Mexican States for the Protection of Migratory Birds and Game Animals. More recently, a Canada-U.S. agreement reached in 1986, the North American Waterfowl Management Plan–Eastern Habitat Joint Venture, was joined by Mexico in 1994. The plan is the basis for several state-provincial action plans (for example, the Eastern Habitat Joint Venture, subscribed to by nine provinces and twenty-one states).

Canada and the U.S. also have signed a wide range of species-specific wildlife protection agreements, including protocols governing the conservation of caribou, polar bears, raccoons, seals, and other species. In 1990, Canada signed its first bilateral agreement with Mexico for the purpose of protecting the habitat of the Monarch butterfly. At the regional level, the U.S. and Mexico both belong to a 1940 agreement, the Convention on Nature Protection and Wildlife Protection in the Western Hemisphere, which established a regional framework for wildlife conservation. The two nations also have several species-specific agreements protecting bats, marine turtles, and others. In 1984, the two countries reached a joint agreement between SEDUE (now SEMARNAP) and the Fish and Wildlife Service (FWS) of the U.S. Department of Agriculture, establishing a Mexico-U.S. Joint Committee on Wildlife Conservation to coordinate bilateral efforts to deal with threatened and endangered species, flora and fauna, exchange of specimens, and migratory bird management. Interestingly, both Mexico and the U.S. have signed the Kingston Protocol to the 1983 Cartagena Convention for the Protection and Development of the Marine Environment of the Wider Caribbean Region. That protocol includes provisions for environmental impact assessment, planning and management, buffer zones, and species recovery plans. It specifically lists a wide range of species of fauna and flora and commits the parties to utilize them "on a rational and sustainable basis." To date, however, neither country has ratified the Kingston Protocol.

It should be noted that a substantial part of the North American conservation effort is driven by the practical need to protect agriculture and property from various pests and blights, and from natural disasters that threaten production and industrial operations. The U.S. and Canada have agreements to combat zebra mussels, inspect plants entering Alaska, and to share pest information for forest management, to include the spruce budworm and other pests. The U.S. and Mexico, similarly, have bilateral agreements to combat equine encephalitis, boll weevil, and the Mediterranean fruit fly. All three countries cooperate on a bilateral basis to prevent and combat forest fires.

Other conservation efforts are associated with the development of parks and protected areas. On the Canada-U.S. border, the International Peace Park, Roosevelt-Campobello International Park, the contiguous Glacier-Waterton national parks, and other protected areas entail some level of international cooperation for their management. On the U.S.-Mexico border, where 18 percent of the border transects some type of protected area in one or both countries, the U.S. National Park Service and SEMARNAP since 1988 have had an agreement on Cooperation in Management and Protection of National Parks and Other Protected Natural and Cultural Sites. Big Bend National Park and Santa Elena Canyon Flora and Fauna Protected Area have received a good deal of attention, including a long-standing informal cooperative arrangement between U.S. and Mexican park managers to manage the systems. More recently, other reserves, including Organ Pipe and Sierra de Pinacate, have gained greater bilateral attention.

Wetlands issues also have received greater attention in recent years. Until fairly recently, wetlands protection was addressed largely indirectly, through the various instruments aimed at protecting migratory waterfowl. In 1988, the three countries signed a Memorandum of Understanding on Wetlands Conservation, which establishes a trilateral committee composed of representatives of the Canadian Wildlife Service, FWS, and SEMARNAP's Natural Resources Conservation division; the aims are to monitor and carry out joint projects for wetlands conservation in the trinational area. All three countries also subscribe to the 1971 international Convention on Wetlands of International Importance, Especially Waterfowl Habitat.

In addition to these bilateral, trilateral, and international agreements, all three countries are signatories of the Convention on International Trade in Endangered Species. Both Canada and Mexico subscribe to the 1993 Convention on Biological Diversity; the U.S. has not yet signed. The U.S. and Mexico also have signed (1979) an

Agreement on Cooperation to Improve Management of Arid and Semi-arid Lands and Control Desertification.

THE SUBNATIONAL LEVEL

There is a surprisingly extensive web of subnational arrangements, and this web of interactions has implications for the operation of NAFTA and its side accords. While interactions among subnational units have emerged in recent decades as an increasingly important aspect of "international" relations generally, this subject has not often been analyzed by either academics or government workers. Indeed, very little solid empirical information exists regarding, for example, the magnitude, range, nature, and operation of ongoing interactions among Canadian provincial governments, U.S. state governments, and Mexican state governments, or between U.S. local governments and Mexican local governments.

There exist only two studies of a comprehensive nature on the interactions between Canadian provinces and U.S. states, and none of the interactions between Canadian provinces and Mexican states, or between U.S. states and Mexican states. The first of the Canada-U.S. studies was carried out in the mid-1970s by Roger Swanson, then of the School of Advanced International Studies at Johns Hopkins University.[3] A replication of this study was conducted in the mid-1980s by Don Munton of the University of British Columbia; John Kirton of the University of Toronto,[4] who surveyed Canadian provincial government officials; and by Lauren McKinsey, of Montana State University, who surveyed state governments through the National Governors' Association. Both these studies are now seriously out of date. Furthermore, no comprehensive compilation of these state-provincial linkages exists, and no provincial government has a current and complete inventory of the interactions in which its own officials engage. Finally, existing studies of state-provincial linkages do not encompass those with Mexico and Mexican states.

The present project identified a total of 103 agreements for the Canada-U.S. border and 42 for the U.S.-Mexico border, all of them in the areas of environment and natural resources (including energy and agriculture).

The agreements identified were classified according to whether they are "formal" or "informal." Formal agreements are those based on written documents, either jointly signed or unilateral in nature (for example, statutes or cabinet orders). Informal agreements are ones based on implicit understandings or verbal commitments. A review of the database shows that, in the case of the Canada-U.S. border, most of the agreements—indeed, two-thirds of them—are formal, compared with 47.6 percent of those for U.S.-Mexico border. For the Canada-U.S. border, this represents a striking contrast to the results of the Swanson study in the mid-1970s, which found that most state-provincial agreements were of the informal variety.[5] It thus appears that linkages of this kind tended to become formalized over the last two decades. Unfortunately, similar comparisons cannot be made for the U.S.-Mexico border. There are other differences between the agreements at the Canada-U.S. border and those at the U.S.-Mexico border, too. While the agreements for the Canada-U.S. border are established between provinces and states, at the U.S.-Mexico border a number of agreements have been established by local communities. The geographical differences between the two borders explain the importance of local actors in the transboundary agreements at the U.S.-Mexico border, where close to nine million inhabitants live in transboundary communities.

At the Canada-U.S. border, more than half of the agreements are multilateral, in the sense that they involve more than two parties. Only a minority is two-party, or bilateral, in nature. The multilateral agreements include relatively small regional accords, as well as arrangements in which all Canadian provinces, all or most U.S. states, and some Mexican states participate. Of the agreements at the U.S.-Mexico border, the majority is bilateral.

All Canadian provinces are directly and actively involved in forming linkages with U.S. states. Some are involved more heavily than others, however. The province most active in forming agreements with states as of the 1990s is British Columbia, which participates in fully 38 percent of all state-provincial interactions. Ontario and Québec follow, being involved in 32 percent each. Alberta, Saskatchewan, Manitoba, and Nova Scotia each are active in about 20 to 24 percent of the agreements. The other provinces, all eastern ones, are less active. The current study included activities of the two territories. Neither is extremely active, but both are involved in this "provincial" activity.

When the activities of U.S. states are studied, the same picture emerges. All states are involved, but some are more active than others. Washington and Minnesota are the two most active. Washington, often engaged with British Columbia, is a party to thirty-four agreements; Minnesota, often engaged with Ontario and Manitoba, is involved in thirty. New York, Montana, and Oregon each have more than twenty-five state-provincial linkages, while Maine, Vermont, Idaho, and North Dakota come in with more than twenty. Other active states include Massachusetts, New Hampshire, Rhode Island, Wisconsin, and California, with twenty each. The prevalence of border states is obvious here, but the high level of activity by nonborder states such as Wisconsin makes it clear that agreements are not reserved for crossborder neighbors. Least active are the southern states of Mississippi, Alabama, Arkansas, Florida, Georgia, Kentucky, North Carolina, South Carolina, and Tennessee. The agreements in which these states participate are almost entirely multilateral in character; they are ones where all the Canadian provinces are linked with all American states, often through broad associations.

At the U.S-Mexico border, Texas occupies almost half of the total U.S.-Mexico border area. Hence it is no surprise that Texas is also the state most dynamic in terms of state and local agreements. Almost half of the agreements are established in Texas (47 percent). California (26 percent) is the second most dynamic state, due to the number of local agreements in the San Diego–Tijuana area; Arizona follows (14 percent). On the Mexican side of the border, Chihuahua and Baja California are the most active states, due mainly to the circumstances of their major border communities, Ciudad Juárez and Tijuana, the two most populous Mexican border communities.

Diverse ministries play roles in state-provincial, state-state, and local-local linkages in the two border areas. Most heavily involved in the agreements considered here, not surprisingly, are the respective environment ministries, followed closely by ministries of natural resources. This is not, it should be noted, a complete listing. Information from the questionnaires sent to provincial, state, and local governments was particularly incomplete with respect to the units of the governments with which arrangements were in place. Provincial, state, and local governments are by no means interacting alone in these agreements. A host of other actors is involved as well. Among these are federal governments and their agencies, municipal governments, ENGOs, private sector interests, aboriginal peoples, and a variety of local and citizen groups.

The respective federal governments, or parts thereof, participate in more than half (52 percent) of the state-provincial linkages on the Canada-U.S. border, and close to 47 percent on the U.S.-Mexico border. In some cases, the states, provinces, and local communities interact within the context of Canada-U.S. or U.S.-Mexico international arrangements. More often, federal entities participate, sometimes merely as observers, in linkages formed by states and provinces.

In some associations, subnational government officials, as well as their federal counterparts, participate. These groups often are as much professional as intergovernmental in character. Examples include the venerable International Association of Fisheries and Wildlife Agencies (formed in 1902), which has members from most states and provinces as well as from the national agencies in Canada, the U.S., and Mexico; and the National Association of State Departments of Agriculture, in which all the American state and Canadian provincial governments have participated since 1978 and in which Mexico has been involved since 1994.

Many of the arrangements provide for participation by other governmental bodies and nongovernmental groups of various kinds. Several ENGOs are involved directly in the agreements, as well as private-sector companies and, in some cases, aboriginal groups. Other organizations of various kinds (universities, professional associations, etc.) participate directly in some of the agreements. Numerous agreements have participation by a variety of groups in more than one of these categories.

There can be no doubt that the current database is incomplete. First, interviews carried out with public officials revealed ongoing initiatives to establish new links that were not covered by the mail questionnaire. For example, during 1995, Alberta and Manitoba were in the throes of negotiating important—indeed, pioneering—agreements with Mexican state governments concerning transportation and other issues. At the present time, these sorts of new initiatives are not included in the database.

The researchers' own knowledge of transboundary cooperation in certain areas suggests that some long-standing interactions are poorly represented in the current database. The Canada-U.S. Great Lakes Water Quality Agreement, for example, has spawned a rather wide range of state-provincial activities, only some of which are documented here. Similarly, the activities of the Eastern Premiers and New England Governors meetings, dating back more than two decades, seem not to be well covered in the database.

Previous research on state-provincial linkages in the Canada-U.S. border strongly suggests that interactions originating in the 1960s, 1970s, and 1980s are not well represented in the current database. In each of the studies done on Canada-U.S. state-provincial linkages (i.e., the Swanson study in the 1970s, the Munton-Kirton study of the 1980s, and the present study), the vast bulk of the interactions documented took place in the time period when the research was carried out. Thus, for instance, the 1980s study found that almost half of the 670 interactions identified and documented (298 in all) originated in the five years prior to that study. Yet the current survey suggests that only 34 interactions originated during that 1981–86 period. While some of these linkages well may have ceased to function, it is highly unlikely that almost 90 percent have lapsed. It is much more likely that they have, through mere "old age" or disuse (but not abrogation), become less well known to the current group of officials who responded to the questionnaire. It is thus likely that a detailed comparison of the results of the current database with the information from the two previous studies, combined with further information from governments, would reveal a substantial number of arrangements that were not picked up by the 1995 questionnaire. This scenario is less likely at the U.S.-Mexico border, where there has been less transboundary cooperation on environmental issues and where the majority of the agreements are very recent (less than five years old).

For all these reasons, the available database cannot be considered exhaustive and needs to be extended, with the aim of compiling a more comprehensive and more complete record of transboundary linkages in North America.

FINDINGS

First, transboundary environmental management is and continues to be highly fragmented in nature. Our review of current institutions and instruments reveals a wealth of transboundary agreements and activities that, all too frequently, are dissociated in organizational and substantive terms. Indeed, the evolution of transboundary environmental management mechanisms has proceeded largely on an ad hoc and functionalist basis, tackling discrete problems either formally or informally as these were recognized at various times and at different levels of government. The few comprehensive, or near comprehensive, instruments currently in place (such as the IJC's formal jurisdiction or the La Paz Agreement, for instance) function as bilateral frameworks for an ad hoc process of cooperation rather than as integrated mechanisms for comprehensive environmental management. The need for greater coordination and integration of multiple disparate management activities within the region is thus apparent.

Second, it is evident that most activity still occurs on a bilateral basis at the several levels of government—national, state-provincial and local. Few trilateral arrangements currently exist in the North American region. Such regional trilateral arrangements as are found— the Memorandum of Understanding on Wetlands Conservation or the North American Waterfowl Management Plan, say—function more as coordinating mechanisms among the three governments and lack the benefit of a common secretariat and other resources for effective trinational management. While the three governments also subscribe to various multilateral protocols, the instruments to which they are party presently do not either adequately extend to, or articulate, the range of environmental problems that could be fruitfully addressed through improved trinational cooperation.

Third, there is ample evidence that, where transboundary environmental management has been extended or deepened, it is presently driven by other multilateral or bilateral agencies. On the U.S.-Canadian border, the IJC continues to play a leading role in the development of new transboundary environmental policy. On the U.S.-Mexico border, a number of new initiatives have been generated by the Pan American Health Organization and by the domestic environmental agencies and natural resource agencies of the two governments. That such agencies are striving to fill recognized voids in the structure of transboundary environmental management is certainly desirable, but it is also evidence of the need for a more effective collaborative mechanism for developing needed transboundary environmental initiatives within the region.

Fourth, cooperation among the national governments revolves around a limited number of agreements: two treaties between the U.S. and each of the other countries concerned with boundary waters; U.S. agreements with Canada on air pollution and hazardous wastes and the La Paz Agreement with Mexico; and agreements concerning wildlife. In addition, the three countries are parties to numerous multilateral environmental agreements that include several or many additional members; in some instances, two of the three countries are parties, even though all three might be expected to be.

Fifth, transboundary state-state, state-province, and local-local interactions seem to have come about by happenstance rather than by design. They have arisen in response to specific historical situations (e.g., an environmental problem requiring coordination) or as the result of specific initiatives by particular state and provincial officials (e.g., a premier pursuing closer political links with neighboring states). Occurring less often are those arrangements between states and provinces and localities that result from international agreements (e.g., the Great Lakes Water Quality Agreement, the International Agreement on Polar Bears, or the La Paz Agreement). Beyond that, a broad and apparently deliberate strategy of developing cooperative links appears only in the case of Québec. To say that there has been no masterplan behind the development of subnational linkages in general is thus something of an understatement. Nor has there been any visible effort to coordinate those interactions that have developed.

There are, in consequence, some clear anomalies in the pattern of state-state and state-provincial interactions, areas where interactions clearly might exist but apparently do not. For example, there is a

Pacific West Coast Joint Oil Spill Task Force dating back to 1989 and involving British Columbia, Alaska, Washington, Oregon, and California. The task force provides for joint emergency response and evaluates prevention measures, capabilities, and technologies for responding to an oil spill. This body does not include Mexico's Baja California, however. While a joint-contingency arrangement exists for dealing with oil spills in the Gulf of Mexico, there does not appear to be an equivalent body on the eastern coast of North America involving the Canadian Atlantic provinces and the Eastern Seaboard states of the U.S. A somewhat similar arrangement was put in place for the Great Lakes some years ago, but it does not appear in the current database and may have become inoperative.

Environmental hazards monitoring is another problem common to all subnational jurisdictions. It is, moreover, one in which cooperation has proven problematical. Current efforts in environmental monitoring are more patchwork than systematic, let alone comprehensive, and thus potentially are open for cooperation and coordination. For example, in the area of transboundary water pollution, British Columbia and Washington established an Environmental Cooperation Council in 1992 to ensure coordinated action and sharing of information on environmental matters of mutual concern. A Georgia Basin–Puget Sound Task Force, set up in 1993, in part coordinates the monitoring of this shared marine ecosystem. Saskatchewan and Montana cooperate on monitoring of water quality in the Poplar River, while Manitoba acts similarly with North Dakota and Minnesota with respect to the Red River. Since 1972, Ontario has cooperated with eight Great Lakes states in monitoring the water quality of the lakes. On the U.S.-Mexico border, the Texas Natural Resources Conservation Commission has coordinated with several Mexican state, federal, and binational agencies to monitor water quality in the Rio Grande. Further west, the Arizona Department of Health and the Santa Cruz County Environmental Health Department have been active in monitoring water quality on the Santa Cruz River, and the California Water Quality Board has been active in monitoring water quality in the notorious New River. These agencies share their information with interested state and federal agencies in Mexico, but the pattern of coordination appears to be ad hoc.

Monitoring of transboundary air pollution presents a similar picture. British Columbia and Washington State established an Air Management Task Force in 1992 to coordinate action and share information on common air quality matters. Saskatchewan and Montana, along with the Canadian and U.S. federal governments, also monitor air quality at the international border as part of the Poplar River Bilateral Monitoring Agreement of 1979. The airshed in the Detroit-Windsor area, too, is monitored cooperatively under a bilateral agreement. On the U.S.-Mexico border, under Annex V to the La Paz Agreement, a binational air quality monitoring program involving federal, state, and municipal agencies on both sides of the border has been established in the Ciudad Juárez–El Paso area, while Baja California and California state agencies cooperate with federal environmental agencies in monitoring air quality in the Tijuana–San Diego area. As in the water quality arena, many of these efforts apparently proceed independently, with little or no interaction among those involved. While there is no need for, and good reasons arguing against, a single integrated continental monitoring program, these various regional efforts could well prove more useful if they were coordinated to a greater degree than they appear to be at present. Any expansion of the 1991 Canada-U.S. Air Quality Agreement, for example, likely would require broader cooperative monitoring by the various jurisdictions.

Sixth, a number of the ongoing interactions identified in this study potentially would benefit, and perhaps would benefit all parties, from being trilateral. As mentioned above, oil-spill prevention and response on the Pacific Coast seem to be areas in which broadening state-provincial cooperation to encompass Mexico is not only desirable but necessary. Other examples of possible "trilateral" state-province cooperation are wildlife management and air pollution. Western provinces and states (including California, Nevada, Arizona, and New Mexico) currently share information on wildlife through the Western Wildlife Health Cooperative. Similarly, British Columbia cooperates with California, Texas, and various northeastern U.S. states under the North American Clean Air Alliance for Zero-Emission Vehicles (established in 1994) to promote development and early commercialization of near zero-emission vehicles on a "national and international basis." Both of these areas would seem relevant to Mexico.

Despite the preliminary nature of this discussion of the database, the results reveal a rich and complex set of transboundary environmental management activities occurring throughout the North American region. These activities create an infrastructure that is useful for the expansion, coordination, and improvement of transboundary cooperation on environmental issues. They are important channels of communication that allow the participants opportunities to exchange views and opinions on common environmental problems and on alternatives for their control and eventual solution. The activities also provide opportunities for the participants in the transboundary activities to establish personal and professional relationships that enhance understanding of, and respect for, their counterparts.

NOTES

1. The opinions expressed by the authors are not necessarily those of the CEC.
2. The CEC has adopted a low-profile role that is essentially additive and does not duplicate the work of the bilateral commissions. The CEC seeks to serve as a forum and a link for coordinating discussion of transboundary environmental issues related to the activities of the bilateral institutions currently in place. It acts as a source of information, including the available database, upon which its various constituencies may draw, and serves as an institutional forum for a range of trilateral environmental activities already conducted by the governments, some of which are related to the mandates and functions of the bilateral institutions.
3. Roger Frank Swanson, *State-Provincial Interactions: A Study of Relations between U.S. States and Canadian Provinces, Prepared for the U.S. Department of State* (Washington, D.C.: CANUS Research Institute, 1974) Roger Frank Swanson, "Relations between States and Provinces," *International Perspectives* (Mar.–Apr., 1976): 18–23.
4. Don Munton and John Kirton, *Province-State Interactions Project: Final Report,* Report to the Canadian Dept. of External Affairs and International Trade (Ottawa: 1988). Don Munton, "The U.S. and Canada: The Linkages of State and Provinces," in *Proceedings of a Conference Sponsored by the 49th Parallel Institute, Montana State Univ., and the Johnson Foundation,* Wingspread Conference Center (Racine, Wisc.: 1988).
5. In the Swanson study, about 30 percent of the interactions were formal, while fully 70 percent were more informal.

Part 1

The North: Canada–United States Case Studies

2

THE BRITISH COLUMBIA–WASHINGTON ENVIRONMENTAL COOPERATION COUNCIL
AN EVOLVING MODEL OF CANADA–UNITED STATES INTERJURISDICTIONAL COOPERATION

JAMIE ALLEY

While there has been a long history of interaction on environmental issues between the Province of British Columbia and the neighboring State of Washington, it is only within the last decade that attempts have been made to develop institutional mechanisms to enhance transborder cooperation. These efforts culminated in the signing of a formal British Columbia–Washington Environmental Cooperation Agreement by the respective heads of government, the premier and the governor, in May, 1992. The agreement established a permanent Environmental Cooperation Council that now has been in operation for five years. The purposes of this chapter are to outline the events that led to the signing of the agreement and the establishment of the council, to review its successes in dealing with transborder environmental issues, and to discuss the lessons that can be learned from this experience.

HISTORY OF THE INITIATIVE

Environmental issues in the Pacific Northwest have a high public profile. As in most areas, watersheds and airsheds in the region do not conform to international borders. There have been formal attempts in the past to resolve issues, such as the Trail Smelter reference (request for investigation and solution) to the International Joint Commission (IJC), in the Columbia Valley in the 1930s, or attempts to cooperate on mutual environmental and economic objectives through formal international mechanisms such as the Columbia River Treaty in the 1960s. More recently, joint environmental reviews, such as the review of the Sage Creek coal mine proposal in southeastern British Columbia by the IJC in the early 1980s, have been conducted. However, much of the ongoing interaction on environmental management issues could be characterized as intermittent and ad hoc. Throughout the 1980s, a whole new set of environmental issues became increasingly important on both sides of the border and had the potential to become major irritants in international relations between Canada and the United States.

On the U.S. side of the border, there was increasing concern over water quality in water courses flowing southward from Canada, particularly the Columbia River, and impacts downstream in Lake Roosevelt in Washington State. There also existed growing concern in Washington over the apparent lack of progress to improve the treatment of sewage discharges to marine waters from large Canadian centers such as Victoria and Greater Vancouver, at a time when cities in Washington were making major investments in sewage treatment facilities in their jurisdictions, as required by U.S. law. Groundwater management also was becoming a major issue, particularly agricultural contamination in the Canadian Fraser Valley of southward flowing aquifers that serve as domestic water sources in neighboring areas in Washington State.

In Canada, concern continued over the transporting by supertankers of Alaskan and other offshore crude oil to refineries in the inner waters of Puget Sound and the catastrophic consequences that a major oil spill could have for the shared marine environment. British Columbians also were concerned about the threat of a nuclear accident, either from reactors in Washington, such as the Hanford Nuclear Reservation and its aging Cold War–era installations, or from the Trident nuclear submarines (based at Bangor in Puget Sound), which traverse the Straits of Juan de Fuca on a regular basis.

On the Nooksack River just south of the border, changes in gravel removal operations and logging in the upper watershed were creating the threat of flooding in Canada, where the potential for major property damage was very high. Canadians also were concerned about the continuing decline of salmon stocks in Washington and Oregon due to overfishing by the Alaska salmon fleet, the resulting pressure on healthier Canadian stocks, and consequent inequities in the allocation of fish under the Pacific Salmon Treaty.

On both sides of the border, general concern was growing over the issue of growth management in the "Cascadia Bioregion," in the face of high levels of net immigration from North America and other parts of the world. The resulting ecological impacts—typified by habitat loss, air and water quality declines in the region, and the loss of marine wetlands—appeared to call for new levels of cooperation in such areas as transportation planning, state-of-environment monitoring and reporting, and strategic planning. The old patterns of intermittent and ad hoc activity, or reliance upon reactive federal mechanisms after problems reached a critical stage, obviously were going to be inadequate to meet the needs of a growing region that places a high value on its natural environment.

This desire to move toward a more proactive model of cooperation was exemplified in the "Speech from the Throne," which opened British Columbia's provincial legislative session in the spring of 1991. It announced a new transborder initiative:

> *Government will take a leadership role in proposing and promoting an international initiative to attack water and air pollution in the entire coastal waterway embracing Puget Sound, Strait of Juan de Fuca, Georgia Strait and waters tributary thereto, including the Fraser River.*
>
> *Discussions will be initiated with the federal government for the establishment of an international commission to include representation of both national governments, the Province of British Columbia and the State of Washington.*
>
> *The requested mandate of the commission will include a thorough analysis of existing and potential pollution threats to air and water, leading to recommendations for a proactive environmental protection strategy for implementation by all four jurisdictions.*[1]

While the political initiative of the British Columbian government was blunted by its electoral defeat in the fall of 1991, the "Throne Speech" commitment did act as a catalyst in beginning discussions among government officials to search for new mechanisms for transborder cooperation. In May, 1991, officials of the Ministry of Environment (referred to here as "Environment" or "the ministry"), met informally with their counterparts from the Washington Department of Ecology ("Ecology" or "the department"), to establish contact and to explore options for enhancing cooperation on environmental issues.

These initial discussions revealed a degree of nervousness by Washington State officials over the concept, outlined in the "Throne Speech," of a "commission" involving the federal governments as equal partners. The department preferred an approach modeled on the successful States–British Columbia Oil Spill Task Force, whereby the province and the state would take the lead in issue identification, analysis, and, where possible, resolution.[2] Only those issues beyond provincial-state jurisdiction would be referred on to the respective federal governments for attention.

These initial informal discussions also revealed a number of important points that colored the final outcome of the initiative:

- Transborder environmental issues were highly charged politically and publicly on both sides of the border.
- Each of the two parties was largely unfamiliar with the roles and responsibilities of the other.
- A mechanism for regular liaison was missing.
- While activities in British Columbia that impact Washington appeared to be more numerous and of greater concern in the short run, the situation had the potential to reverse.

On May 29, 1991, the executives of the Ministry of Environment and the Department of Ecology met formally to discuss the possibilities for moving forward and achieving higher levels of interjurisdictional cooperation. Given the lack of familiarity between the governments, the meeting began with each agency providing an overview of its structure, mandate, and legislated responsibilities and how these fit into the overall operation of the respective governments. It was noted that, although the mandates generally are parallel (Environment has responsibility for fisheries and wildlife, while Ecology does not),[3] there were significant differences in approaches to program delivery, legislation, regulation development, and relations with the federal government that would pose significant challenges if closer cooperation were to be achieved.[4] Agencies also took the opportunity to outline strategic priorities for the next one to three years.

The second half of the meeting was used to review the outstanding transborder issues between the two jurisdictions and jointly to determine their relative significance and priority for action. Issues discussed included:

- Air and marine water quality in Puget Sound and the Georgia Basin.
- Oil supertanker traffic in the Straits of Juan de Fuca and Puget Sound.
- Water quality issues in the Columbia River and Lake Roosevelt.
- Flooding concerns on the Nooksack River.
- Water management, including surface and groundwater issues.
- Solid, hazardous, and biomedical waste shipments.

It became clear from the review of issues that in the past there had been a general lack of communication among officials of the two governments and that many issues had escalated unnecessarily to a high level of public visibility. At the same time, officials of counterpart departments in the two governments were unsure of the degree to which they were mandated or authorized to exchange information or to cooperate with a "foreign" government. As a result, contacts had been irregular, and considerable mistrust had built up. Where officials were working together, they did not have formal reporting mechanisms or means to resolve minor issues and disputes as they arose. Most important, consideration of issues failed to include an understanding of their broader regional context, and no sense of strategic priorities existed. While some first tentative steps were being taken on most of these issues, more formal direction clearly was needed to insure that effective action was taken.

Officials left the meeting with a better sense of each agency's roles and responsibilities, the transboundary issues of concern, and, to some extent, the relative overall importance of each. It was also clear from the discussions that an "institutional gap" existed in the Pacific Northwest; to guide environmental cooperation, an appropriate administrative mechanism was needed. While officials were engaged at the issue-specific level in unmandated and unfunded ad hoc activity on current transborder irritants, the objective often appeared to be to "manage" the issue from a political perspective, rather than to resolve it. Where issues escalated beyond the capacity of regional officials to handle them, often it seemed that the only recourse was to call for a formal review by the IJC. It also could be argued that in the past there had been considerable reluctance on the part of the subnational governments to use the IJC, given that, once a reference was made, they might lose "control" of the issue to their federal governments. Moreover, it might take several years to achieve an outcome, and the process might be costly and time-consuming for the subnational officials.

What clearly was required was a "middle-level" mechanism to fill this institutional gap and see that issues were dealt with in a timely manner. Such a mechanism could be used to share information, foster communication, determine an appropriate strategy to respond to each issue, coordinate activities, and insure action. At the meeting it was agreed to appoint a "negotiator" from each government to develop a proposal for the administrative head of each agency.[5] It was also agreed that any proposal must incorporate the following principles as the basis for future transboundary environmental cooperation: high-level political endorsement, a mandate for sharing information, flexibility in operating style, accountability, and longevity. Consideration also was to be given to the role of the respective federal governments, consultation mechanisms with public and interest groups, and the need to accommodate both the "Throne Speech" focus on British

Columbia's Georgia Basin and Washington State's interest in other transboundary issues of concern, such as the Columbia River.

Despite the lack of precedent for creating this new mechanism, the negotiation process was relatively simple. Several options were considered before the final agreements took shape. These ranged from an international commission as envisioned in the "Throne Speech" to a task force similar to the States–British Columbia Oil Spill Task Force. A commission was rejected due to fears that it might duplicate the work of the IJC, create unnecessary requirements for federal involvement, and be too costly. Similarly, the task force approach was rejected as lacking permanency, being too issue-focused, and lacking in public prominence. This then led to the concept of a permanent British Columbia–Washington Environmental Cooperation Council (ECC), established through a memorandum of understanding signed by premier and governor.

In drafting the memorandum of understanding, the negotiators drew upon the experiences of the Canadian Council of Ministers of the Environment (CCME), which recently had adopted a "Statement of Interjurisdictional Cooperation on Environmental Matters."[6] The CCME statement defines the rationale and nature of interjurisdictional cooperation, provides nine principles to guide cooperative action, and sets out a series of objectives. The statement, which provided particularly helpful guidance to the negotiators, had been prompted by a series of interjurisdictional conflicts among Canadian provinces and with the federal government that in many ways resembled the tensions between British Columbia and Washington.

In addition, the negotiators were guided by the January, 1989, British Columbia–Washington Cooperation Agreement (also known as the Pacific Northwest Economic Partnership), which had been developed to enhance relations between the two jurisdictions in the areas of regional trade, tourism, and investment and to increase economic cooperation. While that agreement anticipated subsidiary agreements (one on energy issues was completed in April, 1989), it was decided that environmental issues were of sufficient importance that they would require a separate "stand-alone" agreement between the heads of government.

The final agreement was composed of three documents: the "Environmental Cooperation Agreement" itself; the "Terms of Reference for the Environmental Cooperation Council," to be established pursuant to the agreement; and a "Preliminary Action Plan and Statement of Work Priorities," which would establish the initial priorities for the council. Although the negotiation of the agreement was relatively straightforward and was largely completed by early fall of 1991, due to the impending election in British Columbia, signing was delayed until May, 1992, when the new government was firmly established and had become familiar with the initiative.

The new premier, Mike Harcourt, while serving for three terms as mayor of Vancouver and during his term as leader of the opposition, already had expressed a strong interest in enhancing regional cooperation. In the final stages of negotiation, both the premier and Booth Gardner, then governor of Washington, took a personal interest in the action plan and work priorities, and directed that additional items be added. The premier and governor had known each other personally for over a decade, dating back to the premier's three terms as mayor of Vancouver and the governor's time as the King County (Seattle) commissioner. The warmth of this relationship was evident at the final signing, which took place in Olympia, Washington, on May 7, 1992.

The agreement's initial clauses set out the two governments' shared commitments to environmental quality, noted the realities of transboundary impacts, and affirmed the desire for cooperative action. Next the document authorized "the Initiative" (later to become the ECC) and committed the parties to the development of the action plan, which was seen as a part of the agreement. Finally, the agreement authorized adoption of specific arrangements to address environmental problems.

The Terms of Reference for the Council, reproduced below, incorporated a number of important features.

British Columbia–Washington Environmental Initiative
Terms of Reference

MANDATE/PURPOSE

The Initiative's mandate is derived from the Environmental Cooperation Agreement between the two jurisdictions entered into May 7, 1992. The Initiative's purpose is to insure coordinated action and information sharing on environmental matters of mutual concern.

MEMBERS

Deputy Minister, British Columbia Environment, Lands and Parks
Director, Washington Department of Ecology

OBSERVERS

Regional Director General, Conservation and Protection, Pacific and Yukon Region, Environment Canada
Administrator, Region 10, U.S. Environmental Protection Agency

SUPPORT

Administrative support will be provided by British Columbia Environment, Lands and Parks and the Washington Department of Ecology, who will be jointly responsible to prepare agendas, insure appropriate attendance at meetings of the Initiative and coordinate followup action.

PROCEDURES
- *The Initiative will generally meet twice each year, or as necessary.*
- *The Initiative may establish subcommittees to deal with specific matters.*
- *The Initiative may, by formal agreement, establish Task Forces to address issues of special or major significance.*
- *An Annual Report will be made to the Premier of British Columbia and the Governor of Washington.*

The council's purposes are two: to coordinate action on outstanding issues and to improve communication between the parties. The precedent of the States–British Columbia Oil Spill Task Force is followed in assigning the federal governments' regional heads with "observer" status. The key organizational decisions—to make the heads of the state and provincial agencies the only formal members of the council, and to assign the federal representatives what would appear to be a subsidiary status—were made after consultation with federal representatives. In reality, this arrangement was adopted largely for practical purposes, to insure that control over the initiative was maintained at the regional level, and to see that the involvement of the federal governments was restricted to areas where they had specific jurisdiction. It also was determined that no new administrative body would be set up to support the work of the council; rather, existing staff would be used.

The Preliminary Action Plan and Statement of Work Priorities noted five high-priority issues for immediate attention: Georgia Basin–Puget Sound water quality, Columbia River–Lake Roosevelt water quality, Nooksack River flooding, regional air quality management, and coordinated groundwater management (Sumas Aquifer). Two emerging issues—solid, hazardous, and biomedical waste cooperation; and water resource management—also were included in the action plan, as well as several other matters of mutual and ongoing interest, such as wetlands preservation and environmental monitoring and reporting. The parties also agreed to use the council to assist each other in dealing with agencies and departments of their respective governments in related matters, such as growth management and earthquake and emergency preparedness.

Since the initial meeting on October 1, 1992, in Seattle, a total of eight meetings of the council have been held. Following each meeting, the deputy minister of Environment and the director of Ecology have made a joint report to the premier and governor. In addition, as required by the agreement, four annual reports have been made to the governor and premier; these have had wide public distribution. Each meeting of the council has been attended by the regional director of Environment Canada and the administrator of EPA (or their designated replacements), representatives of the premier's and governor's offices, staff from the Canadian and U.S. consulates in Seattle and Vancouver, and other invited observers, often from local governments concerned with issues at hand. The meetings are open to the public and normally have attracted a number of members of environmental and other public interest groups. The "Save Georgia Strait Alliance" of British Columbia has been particularly active, as has the American group People for Puget Sound.

The first set of meetings has followed the pattern of joint presentations from staff or invited experts on each of the priority issues, succeeded by a discussion of what is required to move forward on the issue. Opportunities for public comment are provided either after each issue discussion or at the beginning, middle, and end of the meeting. An effort also has been made to set aside time at each meeting for discussion of emerging issues or areas of mutual interest where joint activities may be beneficial.

The ECC has used the model of an international task force for each of the five high-priority issues. In some cases, such as the Nooksack River, where a task force existed prior to the establishment of the council, its members found it useful to be brought under the umbrella of the council. Such status made their work official, gave them a place to report to and resolve issues, and afforded them a direct line to seek funding as required. Given that the council itself reports frequently to the premier and governor, such affiliation gave members added assurance that their work would be recognized. The regular meetings of the council also have served to insure a higher degree of accountability on the part of officials, who know that they will be called upon to make joint reports to the council, outlining their progress or the reasons for the lack thereof. The higher public profile of the task forces also was intended to assure local governments that their concerns were being taken seriously by senior governments.

By the time of its third annual report to the parties in September, 1995, the council was in a position to report significant progress on a number of issues, along with "a general level of improvement in communications between our two governments." The council also signaled that, after the first two years of activity, "the fruits of those labors are beginning to become evident as bilateral teams bring recommendations and action plans back to the council," and that the council was moving "into a new phase of activities."[7]

It is significant that the council also appears to have fostered or encouraged cooperative activity in other areas beyond their own mandate. For example, as the work of the council progressed, the whole issue of growth management became a recurring theme in their activities. The reports of the council were one of the factors that led to the signing, in September, 1994, of a formal "Growth Management Accord" between the two governments. The work of the council, particularly in the area of marine water quality, has also served as a stimulus to independent scientists to establish lines of communication and to work more closely with regulatory agencies.

At a more modest level, the council has been used, as envisioned in the original initiative, to initiate communications among other agencies, such as agriculture agencies responsible for the biological pesticide *Bt*.

By the time of the fourth report, much of the work of the Council and its task forces was moving into a phase of consolidation and institutionalization. The original 1992 agreement had provided for the parties to "enter into specific arrangements necessary to address environmental problems." In April of 1996, the Department of Ecology and the Ministry of Environment, Lands and Parks concluded a more detailed Memorandum of Understanding to clarify roles and responsibilities of their agencies, make specific operational commitments on liaison and communications, and to provide a framework for agreements on specific issues. This led to the development by staff of appendixes on the Columbia Basin and the Abbotsford–Sumas aquifer, which when combined with a 1994 working agreement on air quality, are laying the groundwork for a permanent institutionalization of the goals and objectives envisioned by the original signatories.

SIGNIFICANT SUCCESSES IN TRANSBORDER ENVIRONMENTAL COOPERATION

To illustrate the relative success of the ECC in enhancing cooperation and insuring action on environmental issues, three illustrative examples have been chosen. While these examples are not discussed in great detail, they do give a sense of the nature of the work of the council.

1. Marine Water Quality

No issue facing the council was as politically charged or aroused as much public acrimony as that of marine water quality, and probably no issue has seen as much apparent progress. This also arguably was the single issue most important in the establishment of the council. By the early 1990s, politicians, the media, and community leaders on both sides of the border were outdoing each other in pointing fingers of blame for a perceived decline in marine water quality. Secondary sewage treatment had been installed in Washington under threat of lawsuit by the EPA, and the lack of similar action in Canada was much resented. Business leaders in Seattle and Puget Sound began to organize a tourism boycott of Victoria to protest the slow pace of change in the Capital Region and Greater Vancouver (despite the fact that governments had met with considerable success in joint tourism marketing of the Pacific Northwest). Some American yachtsman were lobbied to stay away from the prestigious annual Swiftsure Classic yacht race. A U.S. senator even wrote to the premier protesting the Canadian sewage from Victoria "washing up on the shores of Washington," despite the fact that scientific monitoring had failed to detect any impacts much beyond the immediate vicinity of Victoria's outfalls, let alone thirty kilometers across the straits in Port Angeles, the nearest American community.

On the Canadian side, officials and politicians pointed to the liquid-waste management planning processes under way in British Columbia and, despite the slow pace of deliberations, warned the other side not to interfere in legitimate democratic processes in a sovereign country. Canadians voiced equal, if less vocal, resentment that American supertankers continued to transit inland waters, delivering heavy crude oil to several refineries; one of these was located at Cherry Point, only a few kilometers south of the border. Canadian environmentalists also were very quick to protest each "friendly" visit to Canadian ports by American nuclear-powered warships, noting allegedly catastrophic threats posed by a potential nuclear accident.

Against this backdrop, the council organized an initial series of information exchanges among agencies and reviewed the work of the Puget Sound Water Quality Authority and the Fraser Basin Management Program. As work progressed, the focus quickly shifted away from the Victoria sewage issue to a greater focus on maintaining high standards of water quality throughout the Georgia Basin and Puget Sound. At their first meeting, in April, 1993, Premier Harcourt and newly installed Gov. Mike Lowry received the council's report on the subject and agreed to establish a joint panel of Canadian and U.S. scientists to examine water quality issues and monitoring under the direction of the ECC. This decision was termed "an important first step in developing a common understanding of the technical information on discharges into the Georgia Basin and Puget Sound."[8] The council was directed to consider nominees and a detailed work plan for the new panel at its meeting the following month.

The International Marine Science Panel was appointed in July, 1993, and charged with giving the council an independent scientific assessment of then-current conditions and trends in the shared waters. Six questions were posed by the council to the panel. The panel in turn organized a major international symposium in Vancouver in January, 1994. The panel invited scientific experts from the region to make presentations on fourteen topical areas, to provide the basis for responding to the council's questions. The presentations were authored jointly by scientists from British Columbia and Washington, and additional participants were invited to the symposium to pose questions to the presenters and the panel. The papers presented and answers to questions from the floor were published in a proceedings document, and, after some deliberation, the panel itself presented its draft findings to the council at its meeting in May, 1994. The full report, "The Shared Marine Waters of British Columbia and Washington," subsequently was released jointly by the premier and governor in September.

The panel presented two possible scenarios for ecosystem health in the next twenty years. The first was a status quo scenario: what would occur if no action were taken. The second scenario outlined the possible state if appropriate management measures were implemented. Given the huge population projections for the region, the status quo projection was very pessimistic and gave the panel's recommendations particular urgency. The recommendations were based on the panel's scientific evaluation of risks and were ranked on the basis of irreversibility, degree of harm, likelihood of prevention, and cost of repair. Their high-priority recommendations, in ranked order, were:

- Preventing estuarine and wetlands habitat destruction and degradation.
- Preventing major freshwater diversions.
- Preventing fish and shellfish population losses.
- Designating marine protected areas.
- Preventing exotic species introduction.

Medium-priority recommendations included:

- Minimizing toxic contamination of marine sediments and biota.
- Preventing large oil spills.

In addressing the domestic sewage issue, the panel noted that, while nutrient enrichment or dissolved contaminants are not a widespread problem, they may be significant in localized settings. Based on a model of currents in shared waters, they concluded that current discharges from Victoria have a negligible effect on Washington water quality.

Having shifted the focus to the regional setting and established a general consensus on the priorities for action, the council was able to move quickly to set up a structure to implement the panel's report through the Puget Sound–Georgia Basin International Task Force. A detailed action plan, including fifty-six possible actions, was presented to the December, 1994, meeting of the ECC and, through a series of working groups, was further refined into specific proposals for consideration at the next meeting in June, 1995. In the implementation of the recommendations the Task Force has made substantial progress in four key areas: habitat loss, marine protected areas, protecting marine life, and minimizing exotic species introductions.

Most significantly, the Task Force has placed a great deal of emphasis on creating a shared, crossborder understanding of issues and approaches through workshops and joint research. Other activities have also focused on education and outreach with citizen groups and the scientific community, and the establishment of a web site on the Internet.[9] While the important work of taking action is only in its early stages, the council's and the panel's efforts have moved this issue from controversy and acrimony to a state in which a shared vision and a common commitment to action are being created.

2. Regional Air Management

At the outset of the council's work, air quality managers in the region had very little contact among themselves, although the general deterioration of air quality, particularly in urban areas, had become a major public concern. With the formation of a task force under the auspices of the ECC, managers began to meet regularly every six months to review and address a variety of crossboundary air quality issues. Initially this work focused on sharing air quality data and information on monitoring networks. In April, 1994, however, it led to the signing, by the Department of Ecology, the Ministry of Environment, and two regional authorities—the Northwest Air Pollution Authority and the Greater Vancouver Regional District—of an "Interagency Agreement" on processes for issuing permits. The agreement provides for prior consultation on major air permit applications that have the potential for transboundary effects within one hundred kilometers, or sixty-three miles, of the border.

Since the inception of the agreement, six "major" and twenty "minor" applications were reviewed. The process appears to be working well and has achieved the desired goal of insuring that regulatory agencies on both sides of the border have input into each other's permitting processes. Equally important, however, are the enhanced exchange by scientists and managers of information on a range of air quality initiatives. In November of 1995, representatives from local air management agencies met with the Air Quality Board of the International Joint Commission to describe the work by the Task Force on air quality issues and initiatives in the Pacific Northwest. The Board was impressed with the activities and indicated that the model may be of use in other jurisdictions.

3. *Columbia River Water Quality*

Discharges into the Columbia River from Cominco Limited's lead and zinc smelters in Trail, and the Celgar pulp and paper mill at Castlegar, British Columbia, historically have been of major concern in Washington, impacting water quality in the river and on Lake Roosevelt downstream from the Canadian border. The Department of Ecology was lobbied strongly in the early 1990s by Washington State citizens to support an IJC reference on this issue. British Columbia's progress in reducing discharges appeared slow and was hampered by technical difficulties. Officials from British Columbia and Washington were very reluctant to support an IJC reference; despite considerable political pressure to make such a reference, they were confident that local solutions could be found in a short time frame. Under the auspices of the ECC, British Columbian officials were authorized to serve as participant observers in the work of the U.S.-based Lake Roosevelt Water Quality Council and to share information on amendments to discharge permits at Cominco and Celgar. In July, 1995, funding for the Lake Roosevelt Council was withdrawn and the original offices in Nelson, B.C., and Spokane, Washington, developed a memorandum of understanding (MOU) to replace the Council's role as a forum for crossborder cooperation.

When communication lines were opened, the atmosphere of mistrust that previously had existed in the area was replaced with a spirit of cooperation and coordination. As a result of improvements to the Celgar mill, levels of dioxins and furans have continued to decline. While high levels of heavy metal contaminants continue to exist in fish and sediments downstream, the construction of new smelting facilities at Cominco, reductions in discharges from sewers, and the ending of slag discharges into the river have resulted in considerable improvements. What is significant in the case of Columbia River is that, while the improvements to Canadian industrial operations were well under way, public concern about the issue did not decline until formal mechanisms for crossborder communication were established, insuring the participation of British Columbia officials in the Washington State administrative mechanisms and vice versa.

IMPORTANT LESSONS FROM THE BRITISH COLUMBIA–WASHINGTON EXPERIENCE

While the evolving model of the ECC is relatively simple and straightforward, its success stands out among recent crossborder experiences, and a number of lessons can be drawn. Of great importance is the fact that the initiative began with the two subnational, rather than the two federal, governments, who pursued it despite initial federal fears of extraterritorial impacts and threats to sovereignty. Too, this may be an indication of growing "bioregionalism" and environmental sensibilities in "Cascadia." Simply put, the citizens of the region are recognizing more and more their common interdependence in environmental matters, despite the existence of an international border.

Eventual federal participation by observers and informal federal approval probably can be attributed to the successful prior experience of the States–British Columbia Oil Spill Task Force, and to clear indications that the initiative would proceed without federal participation in any event. The federal decision tacitly to support the initiative, or the lack of opposition from the Department of External Affairs and the State Department, also reflects the fact that the regional director general for Environment Canada was a former twenty-five-year provincial official who had served as assistant deputy minister in the Ministry of Environment. In addition, the new Clinton administration was very slow to appoint a new regional administrator for Region 10 of the EPA. In short, there was no strong centralist in either federal agency to create alarm in Ottawa or Washington.

It also was very important to the initiative that the mandate began directly with the premier and the governor, and that the council reports directly to them. Never before had environmental issues been given such close attention by heads of government, for such a sustained period of time, and in such a coordinated and strategic fashion. This political endorsement has given the council prominence and legitimacy, contributing greatly to its success. It also has fostered a greater level of commitment at the administrative level, where officials now feel mandated to undertake crossborder initiatives, rather than feeling uncomfortable and sometimes threatened in doing so. For the work of the council to be successful in the future, this level of political commitment must continue. At the same time, it will be important to insure that support for the initiative is not confined to the offices of the premier and the governor, and that at the political level, cabinet and executive colleagues in other ministries and departments are fully engaged and supportive. This will be particularly important in British Columbia, where the Minister of Environment, Lands, and Parks has no clear role in the work of the council and is represented by his deputy minister, who reports on this matter directly to the premier.

The council built upon existing mechanisms and avoided the creation of unnecessary new ones. The direct work of the council and its meetings were funded with existing staff and resources. In contrast to other crossborder initiatives, or the alternative federal dispute-resolution mechanisms, this modest model is logistically appropriate and practical. It should be noted, however, that, the council has no funding mechanisms of its own and so must seek additional resources by redirecting or reprioritizing funds from within existing programs. Given that the work of the council is not directly tied into internal budget and corporate planning processes within each agency, such reallocation may be difficult to accomplish without the support of the program managers involved. It also places the work of the Council in a vulnerable position when agencies face restricted budgets or when substantial downsizing and budget cuts are occurring.

It will be interesting to see if the council can maintain its informal, practical style. Pressures exist to formalize environmental cooperation under the stricter legal requirements of international commitments and conventions or through stipulations by NAFTA and the new North American Commission on Environmental Cooperation. National interest groups often are quite unfamiliar with the regional context or local sensitivities concerning the issues under consideration by the council. Additionally, while the "greening" of trade agreements generally is to be welcomed, the potential involvement of trade litigants in environmental issues may pose some unwelcome challenges.

In that the creation of the council did not include any new mechanism that could impact on or override existing jurisdictional structures and systems of governance, it was forced both to respect them and to work within them. While this necessity presented many practical difficulties, it also imposed a very useful discipline on the participants. Officials in British Columbia and Washington have had to learn about the differences in each other's governments and accommodate varied approaches to resolving problems. This was particularly important with the sewage issue, where very different techniques (standard setting versus site-specific planning) were used to achieve the same objective: a high level of marine water quality in the receiving environment. Meshing very different administrative systems has also presented a variety of problems including such bureaucratically important questions as the equivalency of the official across the table or of the person to whom that official reports. The director of Ecology, for example, is appointed by the governor, while the deputy minister of Environment is the administrative head of that agency but reports to an elected cabinet minister who sits in the legislature.

The council was very successful in depoliticizing a series of very high-profile issues, so that the real public policy issues could be identified and scientific and technical advice examined in a more dispassionate atmosphere. While fortuitous circumstances may have contributed to this outcome, much was attributable to the strategic and contextual framework that the council provided. When seemingly intractable local issues are placed in the broader context of regional priorities and better scientific evidence is brought to bear, varied solutions and outcomes often present themselves.

Finally and most important, the positive benefits of creating and improving communications between the jurisdictions cannot be understated. The council has played a catalytic and high-profile role in fostering improved lines of communication at all levels, ranging from the heads of government to the members of the scientific community, officials in the field, and members of environmental nongovernmental organizations. Almost every issue that the council has faced to date has been clarified greatly, simply by providing a forum for its discussion. Usually the first thing that happened was that participants recognized that they share a common concern and a common objective. Much of this recognition has developed in an informal and unplanned manner. The council sets aside a day for each of its meetings and always includes a meal break, offering a very human opportunity for participants to get to know each other and to build trust. Most participants have had few other opportunities to do this. In addition, the Council has also stimulated increased contact and exchange between academics on both sides of the border as in the case of the Marine Science Panel, as well as between key nongovernment organizations. For example, People for Puget Sound and the Georgia Strait Alliance have banded together on regional issues into the new Sounds and Straits Coalition.

THE FUTURE OF THE COUNCIL

To date, the council has been remarkably successful in its work, attracting the attention of bodies such as the CEC and the IJC as a model of transborder cooperation, but what of the future? In their latest report to the premier and governor, the members raised that question. They suggested that "it may now be appropriate to revisit the original structure and mandate and its relationship to other agencies, levels of government, stakeholders and the public."[10] Certainly, as the first round of issues moves toward resolution and implementation, the council may find itself moving into a less reactive and more anticipatory phase, seeking to resolve problems before they begin. The council also may find itself getting involved in the emerging longer-term issues anticipated in the original work plan, such as state-of-environment reporting and growth management. Long-term success for the council ultimately may rest with expanding their activities to include a higher level of involvement with nongovernmental stakeholders and public interest groups in the region, particularly First Nations in British Columbia and Indian tribes in Washington. Only through the continuing support of the public can the council insure ongoing political endorsement by future governments, a factor that has been critical to its early success.

It should also be noted that the original actors involved in the development of the Council have changed. The new Premier and Governor do not have the personal friendship shared by Booth Gardner and Mike Harcourt, and have not yet established a similar working rapport as their predecessors. The effect on cooperative initiatives of external events in other policy areas also remains unknown, such as British Columbia's highly public campaign throughout 1997 to ensure that the conservation and equity principles of the Pacific Salmon Treaty are fully implemented. Council operations were interrupted at the request of British Columbia in June, 1997, as part of a governmentwide directive to suspend crossborder relations with American counterparts while a cooperative approach and solution to the pacific salmon fishery remains unresolved. In October, 1997, activity at the staff level was renewed, but the Council has yet to resume meeting.

Finally, the Council and the cooperative initiatives sponsored by it may become the victims of their own success. As issues have become depoliticized through scientific and technical cooperation and their public profile as crossborder irritants is diminished, some of the political immediacy that contributed to the Council's early successes may become lost. There is an irony, of course, that as institutional memories of the highly charged conflicts of the past become more distant, and the current impetus for action is reduced, a future generation of actors may have to reinvent some of the cooperative mechanisms that could have prevented another round of crossborder environmental conflict.

NOTES

1. Province of British Columbia, "1991 Speech from the Throne" (Victoria, B.C.: Queen's Printer, 1991).
2. The "States-B.C. Oil Spill Task Force" was created under a Memorandum of Agreement signed by Washington, Oregon, California, Alaska, and British Columbia in 1989, in response to the Nastucca and Exxon Valdes oil spills. Federal agencies are not formal members, and the task force has been used by the subnational governments to ensure that, in addition to coordinating their own efforts, their respective federal governments fulfill their commitments to spill prevention and response.
3. In the government reorganization following the B.C. election of fall, 1991, the ministry also assumed responsibility for crown lands and the provincial park system and was renamed Ministry of Environment, Lands and Parks.
4. An excellent review of federal, provincial, and state government responsibilities can be found in Christine Nasser, "Beyond the Border: Environmental Management in Washington and British Columbia" (EPA 910/9-91-038), a study completed by EPA under the Puget Sound Estuary Program.
5. The author, the chief B.C. negotiator, would like formally to acknowledge and recognize the work of his counterpart in Washington State, Carol Jolly, who at that time was serving as special advisor to the director of the Department of Ecology. In addition, the work of Lynn Bailey and later Glen Okrainetz, who coordinated implementation of the agreement in B.C., also is recognized.
6. The CCME, composed of thirteen federal, provincial, and territorial ministers, is "the major intergovernmental forum in Canada for discussion and joint action on environmental issues of national, international, and global concern." Copies of the statement are available from the CCME Secretariat, 326 Broadway, Suite 400, Winnipeg, Manitoba, Canada R3C 0S5. Publication no. CCME-IC-26E.
7. "Report to the Governor of Washington and Premier of British Columbia from the ECC, Sept. 1, 1995," 6, available from Corporate Policy and Legislation Branch, Ministry of Environment, Lands, and Parks, Parliament Buildings, Victoria, B.C. V8V 1X4.
8. "Joint Statement" by Gov. Lowry and Premier Harcourt, Apr. 5, 1993, available from Corporate Policy and Legislation Branch, Ministry of Environment, Lands, and Parks, Parliament Buildings, Victoria, B.C. V8V 1X4.
9. The web site address for the Puget Sound–Georgia Basin Internatonal Task Force is: http://www.wa.gov/pugetsound/shared/shared.html.
10. "Report to the Governor of Washington and the Premier of British Columbia from the ECC, Sept. 1, 1995," 6.

3

ACHIEVING PROGRESS IN THE GREAT LAKES BASIN ECOSYSTEM AND THE GEORGIA BASIN–PUGET SOUND BIOREGION

R. ANTHONY HODGE
AND PAUL R. WEST

The Great Lakes and Georgia basins are two of five major large-area surface water bodies that occupy parts of the Canada–United States boundary. The complete list includes:

1. The Great Lakes–St. Lawrence bioregion, linking Ontario and Québec with Illinois, Indiana, Maine, Michigan, Minnesota, New Hampshire, New York, Ohio, Pennsylvania, Vermont, and Wisconsin.
2. The Georgia Basin–Puget Sound bioregion, linking British Columbia and Washington.
3. The Gulf of Maine–Bay of Fundy bioregion, linking New Brunswick and Nova Scotia with the state of Maine.
4. Dixon Entrance–Hecate Strait, lying between Alaska and British Columbia.
5. The western Beaufort Sea, linking Alaska and the Yukon.

These transboundary water areas vary greatly in terms of their size, physical characteristics, degree and timing of development, and extent of environmental degradation. However, they share many environmental issues of concern, as well as common federal legislative regimes on each side of the border. Two of these bioregions, the Great Lakes and the Georgia Basin–Puget Sound, are the subject of this chapter. At the core of the discussion, an attempt is made to trace the development that has occurred since Europeans first arrived. This change is assessed from the perspective of achieving progress toward

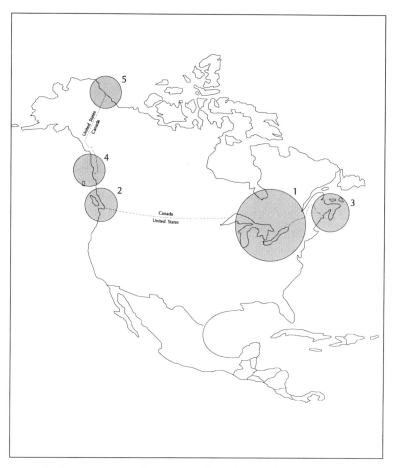

The five major Canada-U.S. transboundary surface water areas.
Map by Clover Point Cartographic Ltd.

sustainability. In short, this means assessing society's success (or not) at improving and maintaining both human well-being and ecosystem well-being jointly. The result is a series of observations and conclusions that are offered as a contribution to the continuing debate about future directions of public policy aimed at addressing transboundary concerns.

Two central themes emerge. First, improving internal federal-provincial/state and a variety of other forms of transboundary cooperation is key to facilitating progress and achieving more effective and efficient action on transboundary concerns. Second, developing a sense of place and community linked to the bioregion greatly enhances the likelihood that action will be taken on bioregional concerns sooner rather than later.

Capacity to assess progress from an integrated perspective of joint human and ecosystem well-being is in its earliest stages of development. In particular, the treatment of human well-being is underdeveloped in this discussion. This limitation is a continuing challenge for emerging systems of results-based management for communities, corporations, and government at all levels.

Of the five transboundary areas, the Great Lakes bioregion has experienced the most extensive industrial and urban development, the greatest environmental degradation, the richest network of transboundary nongovernmental links, and the most evolved institutions of governance. The area emerged as the first industrial heartland of the New World. However, even with the St. Lawrence Seaway bringing international shipping some 3,200 kilometers (2,000 miles) inland from the Atlantic Ocean, the focus of residents to this day remains predominantly continental.

Although much smaller in area and population, the Georgia Basin–Puget Sound bioregion now has a level of per capita productivity that is the same as that of the Great Lakes. This bioregion's marine environment, central role of migratory salmon, focus on the Pacific Rim, lower proportion of heavy industry, and lower overall degree of environmental degradation set it apart from the Great Lakes. However, many of the trends seen earlier in the Great Lakes system—in terms of growth, environmental concerns, shared management of resources with First Nations (aboriginals), and emergence of cooperative transboundary mechanisms inside and outside government—now are appearing here. And, while institutional arrangements for dealing with transboundary issues have some commonalities provided by the respective constitutions, they differ significantly in many ways, particularly in the role that direct province-state relations have played.

These similarities and differences provide the context for undertaking a comparative review. Together they offer a rich fund of experience from which to draw insight. The choice of the two regions of study examined here is merely a starting point. A comprehensive comparative review of all five regions would afford still greater insight and should be undertaken in the future.

In the discussion that follows, the story of each of these areas is sketched briefly. Environmental issues of concern then are summarized and an assessment made of progress achieved to date. Important aspects of governance are described, and the role of developing a sense of place and community within each region is examined. Last, an attempt is made to draw conclusions that might inform policy development in achieving progress toward sustainability.

STORY
Great Lakes Basin Ecosystem

The Great Lakes extend over 1,600 kilometers (1000 miles) along the U.S.-Canada boundary. In addition to the five Great Lakes, the drainage basin includes some eighty thousand smaller lakes. Altogether, the Great Lakes basin ecosystem covers an area of 765,990 square kilometers (295,710 square miles), about a third of which is covered by water. It is shared by eight Great Lakes states (Illinois, Indiana, Michigan, Minnesota, New York, Ohio, Pennsylvania, and Wisconsin) and the province of Ontario.

The following map shows the Great Lakes basin ecosystem, as well as the state and provincial boundaries that serve as a secondary decision-making envelope. Also highlighted is the lower St. Lawrence drainage basin. It would be included in the ecosystem, along with New Hampshire, Maine, Québec, and Vermont, if the entire St. Lawrence drainage basin were to be considered.

Following European contact, the Great Lakes region emerged as the industrial heartland of North America. By 1990, the combined value-added or annual gross state/provincial product of the eight Great Lakes states and Ontario had reached U.S. $1.9 trillion.[1] This figure is roughly twice the annual gross domestic product of the United Kingdom and three times that of Canada. Only Japan (U.S. $2.9 trillion) and the U.S. as a whole (U.S. $5.5 trillion) exceed the amount generated in the Great Lakes region.[2] The brief sketch below describes the changes leading to this remarkable "economic" success.[3]

The first white explorers to see the Great Lakes reached what is now Georgian Bay on Lake Huron around 1615. At the time, the Great Lakes Basin was home to about one hundred thousand First Nations people.[4]

From the time of European contact to the end of the eighteenth century, interest in the fur trade dominated development. Throughout the eighteenth century, British-American military differences impeded settlement. The final fight for the wealth of the Great Lakes region came with the War of 1812. At the end of hostilities, both sides claimed victory. Regardless, the resulting stability, along with an in-place transportation infrastructure and a rich natural resource base, led to wave after wave of settlement and development.

Throughout the 1800s, first the ancient white pines and then the hardwood forests in the Great Lakes region were stripped to clear land for agriculture. To power grist mills constructed to grind wheat and other grains, settlers constructed dams along the thousands of streams and rivers flowing into the Great Lakes. The dams in turn changed the character of the water flowing into the lakes, increasing

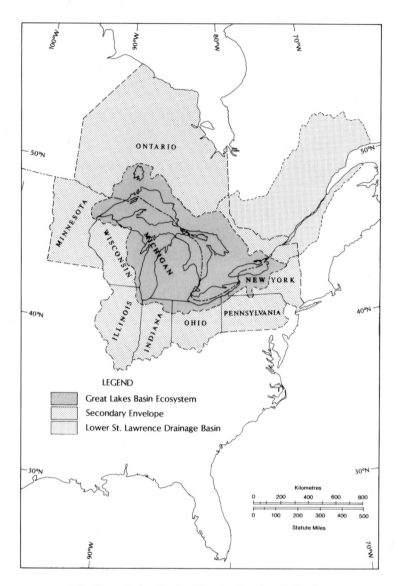

The Great Lakes Basin. Map by Gordon C. Hodge.

water temperature, changing chemical composition, and blocking the migration of river-spawning fish.[5] In time, wood products were sought for markets not only in the U.S. and Canada, but also in Europe. Creeks and rivers were dammed further to provide energy for milling operations, while spring logging drives augmented the damage to river ecosystems. The combined effect of all of the above activities was large-scale and irreversible ecological change.

The harnessing of hydraulic power for electrical generation provided the foundation for the quantum leap in industrial activity that occurred early in the twentieth century. In 1896, with power from Niagara Falls, Buffalo became the first city in the world to be illuminated by alternating current.[6]

Inexpensive hydropower and accessible raw materials drove an iron and steel industry that drew ore from Lake Superior and coal from Pennsylvania. The chemicals industry emerged, with its similar need for both energy and a transportation system to deliver the needed feedstocks and to distribute the resulting products. A broad array of secondary manufacturing and tertiary services proliferated. The transportation and energy support systems grew in sophistication and complexity.

This phase of economic development imposed a second wave of stress on the Great Lakes ecosystem and again had a massive ecological impact. In contrast to the dominantly physical and biological stresses imposed by deforestation, land clearing, and watercourse modification of the first phase, this set of activities generated chemical stresses whose full significance is only now becoming clear. Agricultural chemicals and municipal and industrial waste products were released untreated into the air, rivers, and lakes, or were buried, in the mistaken belief that the subsurface provided safe and stable storage.

Population trends in the Great Lakes region reflect the above changes. By 1900, the population had exploded to ten million from the one hundred thousand found at contact. By 1970, the in-basin population had stabilized at around thirty-five million, roughly twenty-seven million Americans and eight million Canadians. Current estimates suggest a decline on the U.S. side.

There is no doubt that the region's intensive development brought about a spectacular increase in the material standard of living.[7] However, there have been hidden costs—paid for partly in human life but borne mostly by the Great Lakes basin ecosystem itself.

During this century, Great Lakes residents have reacted to no less than five environmental "crises." These crises include:

1. Widespread death from cholera and typhoid at the turn of the century, as a result of drinking water being contaminated by raw sewage.
2. Collapse of the Great Lakes fishery in the 1950s, as a result of competing exotic species, water quality degradation, habitat destruction, and overfishing.
3. Extreme eutrophication in the 1960s and 1970s, as a result of municipal and industrial sewage discharge.
4. Shoreline property damage in the 1980s, as a result of inappropriate land use and record high-water levels combined with severe storm activity.

5. Growing concern since the 1970s with the human and ecological implications of persistent toxic substances that result from in-basin industrial and municipal sources, long-range transport of airborne pollutants from distant sources, and internal cycling within the Great Lakes Ecosystem.

Each of these five "crises" has resulted in unexpected costs to society—costs in human life and health, in a degraded Great Lakes ecosystem, and in damaged public and private property. Only a tiny portion of these costs are factored into the estimates of gross state and provincial product which are used to assess "success" and which identify this region as a major player in the global market. Ironically, it is a quirk of national accounting systems that expenditures to rectify these crises appear as contributions to the growth of state, provincial, and national products.

The Great Lakes region may well be at a critical juncture in its evolution. While expansion continues on the Canadian side, there now are signals that, within the overall basin ecosystem, the continuous population and economic growth of the past two centuries is coming to an end, to be replaced by a stable or even declining population and a more mature steady-state economy. At the same time, there is a growing realization that the hidden costs of success in terms of human health and ecosystem degradation now must be accounted for; reestablishing an enhanced quality of life through ecosystem restoration is emerging as a key to economic renewal. This realization comes at a time when at least one aspect of environmental conditions—concentrations of certain toxic contaminants in the ecosystem—seems to have stabilized at worrisome levels, despite continuing efforts to reduce inputs. From now on, the choices that are made will determine whether or not the region moves into a phase of overall decline or achieves long-term human and ecological well-being.

Georgia Basin–Puget Sound Bioregion

The Georgia Basin–Puget Sound bioregion covers an area of about 135,000 square kilometers (52,000 square miles), half in British Columbia and half in Washington State. The bioregion is an ecological unit bounded by the crests of the Olympic Mountains, Vancouver Island Ranges, the Coast Ranges, and the Cascades. It includes the watersheds that drain into the Strait of Georgia, the Strait of Juan de Fuca, and Puget Sound.

The following map depicts the bioregion, as well as the secondary decision-making envelope that extends to the jurisdictional limits of British Columbia and Washington. Also highlighted is the Fraser River drainage basin in its entirety, only the lower reaches of which are included in the bioregion. The impact of the Fraser, however, is great. It accounts for about 80 percent of the freshwater inflow into the Strait of Georgia and is a major source of land-based inputs to the marine ecosystem. Further, it has played a major role in the history and development of the region.

In the summer of 1792, the English and Spanish arrived simultaneously by sea to map the Georgia Basin–Puget Sound area, fully a century and a half after the Great Lakes were first seen by Europeans. Overland contact from the east occurred shortly thereafter.

The first trading post in the bioregion was established by the Hudson's Bay Company at Fort Langley on the Fraser River in 1827. The economic emphasis on resource utilization that was set at that time continues to the present day: "The northwest economy has always flowed from its natural bounty: abundant salmon and fur-bearing mammals, towering forests, deep soils and mineral lodes, and most recently, the power of cascading water."[8]

Fort Nisqually on Puget Sound followed in 1833. Permanent settlements were established in Victoria in 1843 and Tumwater in Puget Sound in 1845. However, it was the Columbia River gold rush, just to the south in 1855, and subsequently the Fraser River gold rush in 1858 that brought the first massive influx of settlers. While the short-lived gold rushes sparked short-term interest, agriculture, fishing and fish canning, logging, and, in particular, shipping provided longer-term impetus for development.

In time, manufacturing (dominated by pulp and paper and other forest products) gained a central role, but never the heavy industries so important to the Great Lakes economy (with the exception of aluminum manufacturing facilities and the Hanford nuclear facility, which lie outside the bioregion but influence economic activity within it). Military-related activities as a result of World War II, the Korean War, and the Vietnam War played a key role in development, particularly in the U.S. portion of the bioregion. The emergence of the aerospace industry as Washington State's major industry is one result. Recently high-technology and service industries have come to play greater and greater roles in the economies of both British Columbia and Washington.

Links to the Pacific Rim long have been strong. By the mid-1800s, Esquimalt was established as a major supply center for the British Navy. Coal from the Nanaimo coal fields supplied steam ships; then as now, the perspective of residents was as much toward the Pacific as it is inward toward the rest of North America. In this early phase, Great Britain was the major trading partner of British Columbia, while

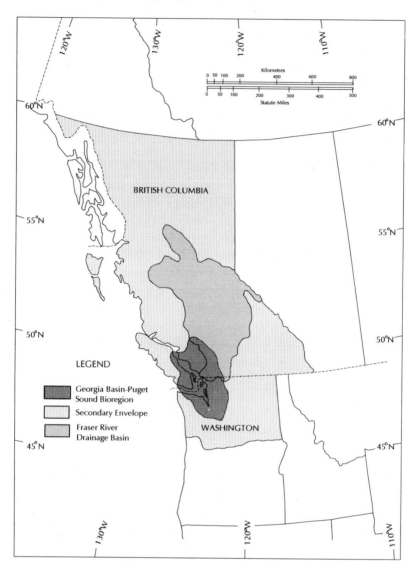

The Georgia Basin–Puget Sound bioregion.
Map by Clover Point Cartographic Ltd.

trade with the U.S. became dominant after World War II. Today, trade with Pacific Rim countries is increasingly important, and both Vancouver and Seattle now serve as major international commercial centers and as gateways to the Pacific Rim.

By 1990, the combined state and provincial annual gross product of Washington and British Columbia had reached $174 billion (in 1990 U.S. dollars).[9] More than three-quarters of this activity is located in the Georgia Basin–Puget Sound bioregion. Interestingly, the per capita productivity of Washington and British Columbia combined in that year was almost exactly the same as that of the eight Great Lakes states plus Ontario. In other words, in a shorter time period, the Georgia Basin–Puget Sound bioregion has achieved the same level of per capita economic success as has the Great Lakes region.

The population of the bioregion[10] has grown from around half a million in the early 1900s to 5.7 million in 1990. Of these people, 2.4 million reside in British Columbia and 3.3 million in Washington state.[11] Since the late 1940s, growth rates have risen steadily and much more rapidly than those of most other areas in Canada and the U.S. For example, between 1970 and 1990, the Seattle region grew by 38 percent, while the amount of land developed grew by a dramatic 87 percent.[12] Through the 1980s, Vancouver and Seattle-Tacoma were tied in having the fourth highest growth rates (12 percent) among all metropolitan areas in the U.S. and Canada with populations greater than one million. (Only Orlando, Sacramento, and San Diego experienced more rapid growth.[13])

Herein lies a major difference between this bioregion and the Great Lakes. In the Georgia Basin–Puget Sound bioregion, rapid growth continues with no end in sight, in stark contrast to the U.S. side of the Great Lakes basin ecosystem, where population is now decreasing. Current projections indicate that the population of the Georgia Basin–Puget Sound Bioregion may almost double by 2015. Urban development could take a range of forms, from vast urban sprawl (like that which has occurred in the Boston-Washington corridor) stretching from Vancouver to Eugene, Oregon, to intensive nodes interconnected by efficient transportation links traversing more rural landscapes. Growth management is justifiably a high priority of both British Columbia and Washington State, since decisions made today will shape the character of their joint future.

The Georgia Basin–Puget Sound bioregion has not experienced a series of identifiable environmental crises, as has the Great Lakes basin. The intensity of heavy industry developed in the Great Lakes never emerged here, population pressures are a more recent phenomenon, and the dynamic nature of the marine environment likely helped it absorb many waste products that would have been more evident in a freshwater lake environment. However, in this region concern is growing about air quality, surface water quality, persistent toxic substances, the current crisis in the fishing industry, and habitat and agricultural land destruction related to urban sprawl. This concern echoes similar concern in the Great Lakes region. Furthermore, evidence is mounting that, in the Georgia Basin–Puget Sound bioregion, counterparts exist to the human and environmental costs of development success that only now are being accounted for in the Great Lakes system. The difference is that the Georgia Basin–Puget Sound bioregion has not been degraded to the extent of the Great Lakes system, and here the choice exists whether or not further degradation is to be avoided as development proceeds. As one observer put it, "Though the region's economy is badly out of balance with the ecosystems it draws upon, its environment is probably less degraded than any populated part of the industrialized world. No place on earth has a better shot at reconciling people and nature than the Pacific Northwest, the greenest corner of history's richest civilization. And with most of the planet's people aspiring to our North American standard of living, no place has a greater responsibility to set new standards for an ecologically endangered world."[14]

ENVIRONMENTAL ISSUES OF CONCERN

Industrial and urban growth of the nature, scale, and rate which has occurred in both the Great Lakes basin ecosystem and the Georgia Basin–Puget Sound bioregion has imposed physical, chemical, and biological stress on the enveloping bioregion. The result is a broad range of concerns that must be dealt with by public policy. The nature and extent of actual damage varies significantly within and between the two regions. However, almost all the specific environmental issues of concern found in one are common to the other. These include:

- Threatened biodiversity—ecosystem, species, and genetic—as a result of habitat degradation, increased toxic substances, and introduction of exotics.
- Decreased air quality and changes to climatic conditions as a result of emissions from both local and far-distant sources. These emissions include toxic contaminants (metals and persistent organic pollutants), common contaminants (carbon monoxide, ground-level ozone, fine particulates, nitrogen oxide), acid contributors (sulphur dioxide, nitrogen dioxide), stratospheric ozone depletors, greenhouse gases, and thermal loading in urban areas.
- Contamination and thermal loading of surface water and groundwater from a variety of point and nonpoint sources.
- Adjustments to surface water flow regimes as a result of diversions, dams (for power and flood control), and excessive withdrawals.
- Depletion of groundwater supplies as a result of excessive withdrawals.
- Land quality degradation from conversion of agricultural land to urban uses, loss of wetlands and forests, soil erosion, depletion of soil productivity, and shoreline erosion.

- Degraded built infrastructure and urban sprawl, leading to significant inefficiencies.
- Loss of environmental quality to the extent that human well-being is affected: increased exposure to pollutants, noise, degraded built infrastructure, urban sprawl leading to significant inefficiencies, less appealing physical and social conditions impede economic development, and an overall reduction in quality of life.

Of these issues and from a purely physical perspective, the "transboundary" elements include biodiversity, air quality, climate, and changes to the surface and groundwater regimes. However, with the signing of the North American Free Trade Agreement (NAFTA) and the resulting pressure to establish a regulatory level playing field across the North American continent, all of these issues to some extent assume a transboundary scope.

PROGRESS ON ENVIRONMENTAL ISSUES

The first assessment of pollution of boundary waters in the Great Lakes was completed in 1912,[15] although initial recommendations for cleanup and prevention of further degradation never were implemented. Additional studies were completed in 1946, 1948, 1964, and 1966.[16] By the late 1960s, Lake Erie was being described as "dying," and massive mats of algae were visible throughout the lower Great Lakes, a sign of serious eutrophication. Perhaps the most dramatic event occurred on June 22, 1969, when the Cuyahoga River in Cleveland, Ohio, was so heavily laden with high concentrations of oil and other flammable industrial wastes that it caught fire and burned two railway bridges beyond use. Finally, in 1972, sixty years after the first study of transboundary pollution, the governments of Canada and the U.S. signed the Great Lakes Water Quality Agreement.

In the following two decades, remarkable success was achieved in reducing nutrient discharges to the Great Lakes and controlling the problem of eutrophication, through construction of industrial and municipal sewage treatment facilities. Over ten billion dollars (1989 U.S. dollars) were spent on municipal sewage treatment facilities alone. Investment in pollution control by private industry likely has been as large. However, a comprehensive review of the state of the environment of the Great Lakes, completed in 1990, concluded: "Despite regulatory vigilance to rein in polluters and significant government cleanup efforts over the last two decades, the environment of the Great Lakes basin is still in trouble. Dramatic evidence remains that the Great Lakes are imperiled by continuing habitat destruction and the long-term accumulation of toxic chemicals, which are increasingly pervasive throughout the ecosystem."[17]

These conclusions were reconfirmed in followup studies completed in 1991[18] and 1995.[19] One particular trend now evident is particularly troubling: although reductions in the concentrations of a number of persistent toxic substances have occurred, the rate is slowing; in the case of Lake Ontario, concentrations appear to be stabilizing at unacceptably high levels.[20] This plateauing effect is not completely understood but likely is due to a combination of (1) continued use in remote areas and subsequent transport into the basin (particularly via the atmosphere); (2) continued discharges from point sources even if at much reduced levels; (3) continued release from improper storage of waste and remaining stocks, often in the form of contaminant migration in groundwater flow systems (hazardous waste sites in particular); (4) continued discharges from nonpoint sources; and (5) bioaccumulation, rerelease, and recirculation within the food web. In addition, reductions in the level of effort put by governments driven by budget-cutting exercises now is causing significant concern. In its 1996 report on Great Lakes water quality, the International Joint Commission on the U.S.-Canada Border (IJC) states:

> *Investments in sewage treatments, stormwater runoff management, controls on industrial discharges, shipping and dredging, and limited bans on phosphorus and certain pesticides have produced significant results. . . . Despite this success . . . the progress of the last quarter-century of investments in the Great Lakes is in jeopardy. The following proposals and actions in both countries place in question their capacity to sustain this progress:*
>
> - *proposals to weaken regulatory frameworks that underpin pollution control and other effective programs, including reporting and compliance requirements;*
> - *erosion of funding and expertise for research, monitoring and enforcement, and transferred responsibilities to other levels of government without the required resources.*[21]

In contrast to the Great Lakes ecosystem, no comprehensive transboundary state-of-environment assessment of the Georgia Basin–Puget Sound bioregion ever has been completed. While such a synthesis has not been made, a number of important pieces have been completed. These include:

- An assessment of the state of the shared marine waters, undertaken for the British Columbia–Washington Environmental Cooperation Council by the British Columbia–Washington Marine Science Board.[22]

- A number of "State of the Sound" assessments completed by the Puget Sound Water Quality Authority.[23]
- State-of-environment reports completed for both Washington[24] and British Columbia.[25]
- A review of urban sustainability, completed by the former British Columbia Roundtable on the Environment and the Economy.[26]
- A comprehensive assessment of progress toward sustainability (including an assessment of ecosystem well-being) that now is in preparation for British Columbia.[27]
- Various regional planning studies focused on Vancouver and Seattle.
- A variety of studies of the lower Fraser, completed as part of the work of the Fraser Basin Management Program.[28]
- An assessment of the state of the entire Pacific Northwest drainage, from northern California to Prince William Sound in Alaska, completed by Northwest Environment Watch.[29]

A very similar picture to that of the Great Lakes emerges from a review of these documents: there have been some successes, and encouraging trends are developing; but overall ecosystem integrity, now and in the future, is far from confirmed. For example, in the Puget Sound area, significant progress has been achieved in upgrading municipal sewage treatment facilities to secondary treatment and reducing attendant discharges of both nutrients and toxic contaminants. In contrast, only in 1995 did the Greater Vancouver Regional District come to a cost-sharing agreement with municipal, federal, and provincial partners to complete upgrading of Vancouver sewage treatment facilities to at least primary treatment. Victoria still discharges raw sewage into the Strait of Juan de Fuca through long marine outfalls, although scientists, engineers, and policy makers remain locked in a debate about the extent of the resulting environmental damage and the benefits to be achieved by putting scarce resources into sewage treatment facilities. In both the Great Lakes and the Georgia Basin–Puget Sound, significant improvements in air quality have been achieved, but urban smog (described, in Chicago, as "toxic soup"[30]) remains a serious problem in many cities, particularly with regard to ground-level ozone. Air quality problems in Vancouver often are compared to those of Los Angeles.

In British Columbia through the past decade, bleach kraft pulp mills have achieved significant success in reducing the levels of total suspended solids, biological oxygen demand, dioxins, and furans in the effluent, through both process changes and the use of secondary treatment systems. Contaminant levels in seabird eggs from nearby sites similarly have been reduced. Heavy metals in biota close to a number of industrial sites (for example, mercury in crab in Howe Sound, in the vicinity of a now-closed mercury-cell chlor-alkali plant) also now register significantly below 1970 levels.[31] As a result, several previously closed fisheries, including the prawn and shrimp fishery in Howe Sound, have been reopened.

As with the Great Lakes, these successes must be viewed in context. For example, concentrations of polyaromatic hydrocarbons, or PAHs (primarily derived from burning of fossil fuels), in sediments in central Puget Sound currently are less than half the peak concentrations from the 1950s, but still are fifteen times higher than the baseline levels from the 1880s. Similarly, concentrations of some contaminants in surface sediments in parts of the Sound's urban bays are elevated one hundred times above the levels of the cleanest rural bays.[32]

This kind of historic comparison raises questions: So what? And just what levels can be considered "safe, clean, or adequate"? While there are no crisp and simple answers to such questions and debate among scientists abounds, the plateauing of levels of toxic substances in parts of the Great Lakes system and growing concern regarding bioaccumulation in the food web and potential human health effects suggest that a high degree of caution is warranted.

While examples in the Great Lakes are more extreme, in both areas there is increasing documentation of reproductive failures, deformities, and physiological malfunctions in fish and wildlife. There is a growing body of evidence linking these problems to a variety of persistent toxic substances, especially those with a capacity to mimic the female hormone, estrogen, and the male counterpart, androgen.[33] And while scientific debate is heated, there is growing published evidence indicating that harm to humans from persistent toxic substances is similar to that caused in wildlife.[34] This set of concerns now is receiving priority treatment by agencies in both the U.S. and Canada.

In both the Great Lakes and the Puget Sound portion of the Georgia Basin–Puget Sound bioregion, over 60 percent of all wetlands have been destroyed since European settlement.[35] Further, continuing conversion of forest and agricultural land to other uses (particularly as a result of urban sprawl) persists in reducing both the habitat available and the quality of the habitat remaining. In the Fraser River delta, tidewater habitat used by juvenile fish and migratory birds is a particular concern.

In summary, while there are large variations in the degree of degradation across and between both regions (ranging from pristine to highly degraded) and significant differences in both the time period and extent of settlement and industrial development, a number of common attributes can be identified. Specifically, both regions are characterized by some degree of progress, long time lags between problem identification and action, and continuing indications that the ecosystem remains stressed and far from healthy.

GOVERNANCE

In both the Great Lakes basin ecosystem and the Georgia Basin–Puget Sound bioregion, drainage basin characteristics can be used to set a primary boundary for analysis. The maps of the Great Lakes Basin and the Georgia Basin–Puget Sound bioregion also show a secondary decision-making envelope that includes the full extent of the implicated states and provinces. This envelope is important to recognize because state and provincial decision makers will use the context of their entire jurisdiction when making decisions about the more limited area within either the Great Lakes basin ecosystem or the Georgia Basin–Puget Sound bioregion. (Michigan is an exception, as the entire state is located within the Great Lakes basin ecosystem.)

In each of the two cases, there is an immensely complex web of institutions with complementary and often overlapping responsibilities that touch on the environmental issues discussed above. For example, the Royal Commission on the Future of the Toronto Waterfront (Crombie Commission) notes that, in the Greater Toronto bioregion alone, more than a hundred agencies in five layers of government share responsibilities for Great Lakes matters. The result is a kind of "jurisdictional fragmentation" that makes pinpointing responsibility difficult at best and in some cases futile.[36] The commission points out that "the existing regulatory framework is characterized by overlap and duplication by different levels of government, by joint action on some issues, and by failure to exercise authority that is already in place. . . . The framework is fragmented, with different instruments governing separate aspects of the environment, which makes it difficult to apply ecosystem goals and principles."[37]

The same could be said of Chicago, Detroit, Seattle, Vancouver, or any other community in the Great Lake basin ecosystem or the Georgia Basin–Puget Sound bioregion. However, despite this maze of institutional players, the basic structure of "environmental" governance for these two areas is relatively simple. The following charts depict that structure for the Great Lake basin ecosystem and the Georgia Basin–Puget Sound bioregion, respectively.[38] Comparing the charts allows us to identify a number of key relationships of governance that are less evolved (in terms of formalized arrangements) in the Georgia Basin–Puget Sound bioregion than in the Great Lakes basin ecosystem. The following observations arise.

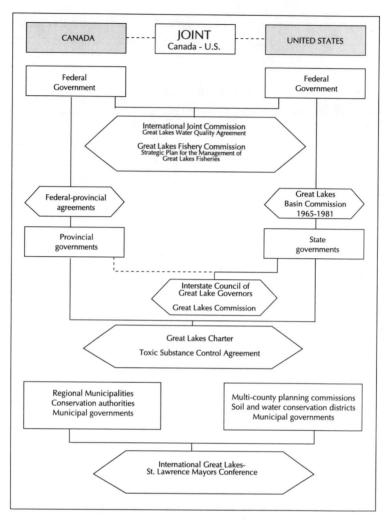

Basic framework of governance for the Great Lakes basin ecosystem. Chart by Clover Point Cartographic Ltd.

Federal-Federal Relationships and the Key Role of the IJC and Its Great Lakes Regional Office

In the Great Lakes, two transboundary federal-federal institutions have been created with a regional focus: the Great Lakes Fisheries Commission (GLFC) and the Great Lakes Office of the IJC. In the Georgia Basin–Puget Sound, the Pacific Salmon Commission is somewhat comparable to the Great Lakes Fisheries Commission, but there is no equivalent to the IJC Great Lakes Office.

The GLFC was established in 1955 by the Convention Between Canada

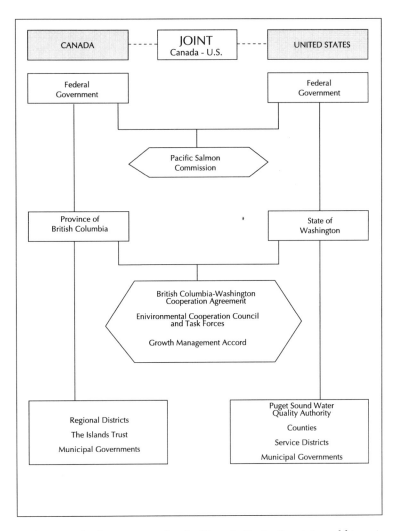

Basic framework of governance for the Georgia Basin–Puget Sound bioregion. Chart by Clover Point Cartographic Ltd.

and the U.S. for the Conservation of Great Lakes Fishery Resources. Efforts to establish international fishery commissions and/or effective, complementary regulations and management programs for Great Lakes fisheries failed repeatedly from 1893 to 1952. By 1946, the sea lamprey, a parasitic predator native to the Atlantic Ocean, was established throughout the Great Lakes, creating an impending catastrophe for the fisheries. With this incentive, negotiations finally were completed. By that time, unfortunately, the commercial lake trout catch from Lakes Huron and Michigan was 99 percent lower than the average annual catch during the 1930s.[39] The GLFC now is charged with overseeing the Strategic Plan for the Management of Great Lakes Fisheries.

In the forty years since it was created, the GLFC has overseen a spectacular redevelopment of the fisheries. Much of this success has been "bioengineered" through control of exotics, stocking programs, and other measures. The Great Lakes fishery now is managed like a giant fish ranch dependent upon introduced fish. It may be the largest such experiment in the history of humankind. In spite of this effort, natural reproduction in balanced fish communities has not been reestablished.[40]

In 1985, after many years of negotiation, the U.S. and Canada signed the Pacific Salmon Treaty. The treaty establishes a framework for cooperative management, research, and enhancement of Pacific Salmon stocks. It embodies commitments to prevent overfishing, achieve optimum production, and ensure a fair distribution of benefits between the two countries. The Pacific Salmon Commission was established to implement the treaty. Its mandate is limited to managing the harvests of sockeye and pink salmon destined for the Fraser River and passing through Juan de Fuca Strait, and dealing with international allocation and protection of chinook, coho, and chum salmon taken in coastal troll and recreational fisheries. It does not deal with domestic allocations within the U.S. and Canada, nor is it mandated to provide advice on overall fisheries management.[41]

Overall management of the fishing industry and monitoring of the aquatic ecosystem have been continuing sources of contention throughout the Pacific Northwest. Through the mid-1990s, continuing depressed fish stocks led Canada to reduce its fishing fleet significantly, an action that had a major impact on many small coastal communities. In the summer of 1997, U.S.-Canada negotiations to renew the Pacific Salmon Treaty collapsed, British Columbia severely curtailed direct relations with the government of Washington on all fronts, and British Columbia's relationships with the Canadian government were seriously impaired. In Prince Rupert, frustrated fishermen blockaded the Alaska ferry, leading to a cessation of service and subsequent court action. Taken together, these "fish wars" and the resulting temporary impairment in cooperation may have been essential to drawing the attention of the U.S. president and the Canadian prime minister to this issue. However, solutions ultimately must emerge from renewed cooperation, not antagonism.

In 1909, Canada and the U.S. signed the Boundary Waters Treaty, establishing a framework of cooperation between the two countries on transboundary water issues. Although original motivation stemmed mostly from water allocation issues related to hydropower development, diversion, irrigation, and flood control, the treaty also enshrined a basic principle that the boundary waters "shall not be polluted on either side to the injury of health or property on the other."[42]

A key element of this framework is the IJC, a binational commission that was established to review issues of concern upon reference by the federal governments. The advice of the IJC, while not binding on the parties, is intended to contribute to good decision making and, when required, assist in dispute resolution.

In 1972, mainly in response to the massive problem of nutrient enrichment (from industrial and municipal sewage) and resulting eutrophication, Canada and the U.S. signed the Great Lakes Water Quality Agreement (GLWQA). Among other terms, the agreement directed the IJC to establish a Great Lakes Regional Office.

In 1978, the agreement was amended by protocol, shifting the focus to include toxic substances and requiring preparation by the IJC of a biennial assessment of Great Lakes water quality. It also committed the parties to a basinwide ecosystemic view in development and implementation of policies aimed at restoring and enhancing Great Lakes water quality.

This commitment is based upon a conceptual foundation that sees people as parts of an enveloping natural system. It demands consideration of, and sensitivity to, the complexity of that system. It shifts the perspective of water quality management from a limited focus on lake water quality pure and simple, to a broad perspective on the interconnected influencing factors that occur both in water and on land that relate to water quality. In application, the commitment backs the IJC into considering the complete Great Lakes basin ecosystem. It does so recognizing that dealing with the water quality issue is possible only through consideration of the larger enveloping ecosystem.[43] This shift occurred because of a recognition that profound changes were required in how people treated the Great Lakes. If the system that includes water quality was to be restored, any human use of the natural resources of the Great Lakes basin ecosystem could be only on a "sustainable" basis, that is, designed to improve or maintain human and ecosystem well-being now and in the future.[44]

Every two years since the signing of the original 1972 Great Lakes Water Quality Agreement, the IJC has met in a public forum around the Great Lakes to review issues related to water quality and hear the views of the public. In 1975, about a dozen citizens participated. Twenty years later, in 1995, some two thousand joined the commission for its three-day biennial meeting. The public interest and participation that these biennial meetings have spawned are critical contributions.

The year 1997 marked the twenty-fifth anniversary of the GLWQA. Various reviews of its effectiveness are currently under way, and there is much debate about lessons from the past and directions for the future.[45] However, in the context of this discussion, it is most important to note that any progress that has been achieved within the Great Lakes basin ecosystem has been greatly facilitated by the technical contributions and integrating efforts of the IJC's Great Lakes Regional Office and its various related boards and task forces.

In the Georgia Basin–Puget Sound bioregion, no federal-federal institutional mechanism exists that is equivalent to the IJC. While transboundary advisory boards and task forces have been set up from time to time at the federal level, they have been short-term and have functioned in the absence of any overall strategic framework. There is considerable resistance in British Columbia and Washington State to creating a mechanism that, like the Great Lakes Office of the IJC, is controlled federally (from twenty-five hundred miles away, on the other side of North America). However, from its creation in 1992 until the suspension of its activities in June, 1997, the British Columbia–Washington Environmental Cooperation Council (ECC; discussed below and in Jamie Alley's chapter in this volume) performed some similar functions, albeit on a much smaller scale in terms of staff, financial resources, overall activity level, provision of a public focus, and public participation.

The above discussion of federal-federal arrangements is not intended to imply that federal leadership is necessary to effect timely solutions to transboundary issues of concern. In the Georgia Basin–Puget Sound bioregion, a great deal of communication and cooperation takes place informally and formally at the province, state, and municipal level. This activity is discussed below. However, what does seem apparent is that the most effective route to solution building is one in which cooperative approaches are generated that link resources, insight, and political will at all levels of government. For a variety of reasons, this seems to have occurred more effectively in the Great Lakes basin ecosystem than in the Georgia Basin–Puget Sound bioregion.

Federal-Provincial/State Relationships

Regulatory and legislative cultures in the U.S. and Canada differ significantly.[46] It is particularly important to understand these differences when considering federal-provincial/state relationships.

In the U.S., regulatory and related action (reflecting desired policy directions) tends to be spelled out in detail in legislation specifying the kinds of regulations to be issued by the executive branch and strict deadlines for their issuance. Much time and energy subsequently is consumed in implementation. Provision for judicial and quasi-judicial review and for public comment is entrenched, and litigation on the part of both government and citizens is the norm.

In contrast, Canadian legislation often is more limited, in the sense of providing only general policy direction and leaving considerable discretion to ministers and their departments in determining the scale, scope, style, and speed of implementation. There tends to be much less opportunity for public review of executive decisions than in the U.S. and even less opportunity for judicial review. The results have been described as leading to a kind of "tyranny" in the British parliamentary system. Some Americans find the discretion and apparent lack of accountability in the Canadian system disconcerting, while the U.S. system appears to some Canadians cumbersome and unnecessarily adversarial. (In 1995, the government of British Columbia moved to enhance the degree of accountability, in an initiative supported jointly by all parties of the legislature, all ministries, and the provincial auditor general. This "Accountability Initiative" has involved developing new systems of performance measurement and reporting, linked to results-based management.[47] Similar efforts to enhance accountability also are being made in other provincial governments and the federal public service.)

Significant differences also exist in the federal-province/state division of powers between the two countries. In Canada, there are some clear areas of federal environmental jurisdiction—for example, as established by the Fisheries Act, the Canada Oceans Act, and the Canadian Environmental Protection Act. However, with both natural resources (with the sole exception of marine fish) and municipal affairs identified by the constitution as provincial responsibilities, provincial governments have primary responsibility for environmental protection and natural resource management. In contrast, the U.S. federal government has broad authority, strongly supported by the courts, to address environmental issues. It is the federal government that defines the major steps that the states and local governments must take to develop and administer environmental programs. While some states (Washington among them) have developed programs that exceed federal requirements, many simply allow their programs to be shaped by federal requirements.

Another major difference is that the U.S. government has the constitutional power to override state authority to insure implementation of international agreements. That is not the case in Canada. For example, before Canada can agree with the U.S. on a strategy to achieve ecosystem health in the Georgia Basin, it must first agree with British Columbia. If the government of British Columbia were unwilling to adopt or implement such a strategy, the federal government could neither compel it to do so nor intervene with a comprehensive set of actions of its own. While Canada has signed NAFTA, application of its provisions is assured only if the provinces included in any application account for at least 55 percent of Canada's gross domestic product. To date, only Alberta, Québec, and Manitoba have signed. In contrast, in the U.S., if a state fails to develop a satisfactory program under the U.S. Clean Water Act to control municipal and industrial discharges to surface waters, the federal government develops and implements such a program.

In spite of these differences, it is apparent that, on both sides of the border, effective federal-state/provincial cooperation and the political will to act at both levels of government are more critical to initiating action than a particular split in constitutional responsibilities. The old adage, "Where there is a will, there is a way," applies here. This conclusion is hard to avoid when comparing the Great Lakes basin ecosystem and Georgia Basin–Puget Sound bioregion.

In the Great Lakes, to facilitate a cooperative effort on the Canadian side, the Canada-Ontario Agreement on Great Lakes Water Quality has been signed by four federal departments and four from Ontario. While federal and provincial positions are not always identical, this agreement signals a significant degree of joint effort. No doubt a key to this situation is the political importance in Canada, to both the provincial and federal governments, of the Great Lakes region, which contains the largest concentration of voters in the country. Regardless of motivation, one result that may be linked to more effective federal-provincial cooperation is that the "Canadian side" often (though not always) has played a leadership role in the overall Great Lakes basin ecosystem.

The opposite appears to be the case in the Georgia Basin–Puget Sound bioregion. There is a long history of antagonism between the governments of western Canada, particularly British Columbia, and the federal government.[48] During the late 1980s and until the collapse of fisheries negotiations in the summer of 1997, several joint federal-provincial initiatives tentatively heralded a new era of collaboration. Among these initiatives were programs aimed at the Fraser River (Fraser River Management Program, Fraser River Action Plan, Fraser River Estuary Management Plan), the Burrard Inlet Environmental Action Plan, and the Victoria Esquimalt Harbour Environmental Action Plan. The British Columbia–Canada Infrastructure Works Program is providing funding for a number of projects, including the upgrading of sewage treatment facilities in the Greater Vancouver Regional District. A "Lower Fraser–Georgia Basin Ecosystem Initiative" currently is being designed cooperatively by Environment Canada, the British Columbia Ministry of Environment, Lands, and Parks, and other federal and provincial partners. And, importantly, in April, 1997, Canada and British Columbia signed the Agreement on the Management of Pacific Salmon, entrenching a cooperative approach on fisheries policy development and management on the British Columbia coast.

However, no formal coordinating mechanism ever has been created for the Georgia Basin that would serve as a Canadian counterpart to the Puget Sound Water Quality Authority. Created in 1985, the latter was assigned the task of developing a comprehensive management plan that would bring some degree of cohesion to "6 federal agencies, 5 state agencies, 12 county governments, 10 city governments, 14 tribal governments, 40 port districts, 13 sewer districts, 15 flood control districts, 121 soil and water conservation districts, 14 parks and recreation districts, and 9 public utility districts."[49] In addition, no umbrella agreement that would provide a common framework for action by both province and federal government entities, analogous to the Canada-Ontario Agreement on Great Lakes Water Quality, ever has been consummated for the Georgia Basin.

While the ECC (which includes participation by federal observers) was created to deal with a range of transboundary environmental concerns, it never enjoyed the stature or support, in terms of resources, that the IJC's Great Lakes Office has enjoyed.

Historically, the Georgia Basin has not had the demographic, political, and economic importance for the Canadian government that the Great Lakes region has. This situation will change as the political and economic strength of British Columbia grows relative to other parts of the country. The growth in importance of Pacific Rim trade undoubtedly is a contributing factor. In any case, until the Harcourt era in provincial government, the "Canadian side" never had played a leadership role in resolving common environmental issues in the Georgia Basin–Puget Sound bioregion. Instead, the standard had been set by Washington State, in cooperation with the federal U.S. government, through the Puget Sound Water Quality Authority.

The reasons underlying this apparent Canada/Ontario leadership in the Great Lakes and the U.S.-Washington leadership in the Georgia Basin–Puget Sound bioregion have never been examined in detail and would form an interesting topic for future research. The answer may be as simple as self-interest: the lead is taken by the population that stands to be most affected by environmental degradation. However, it is difficult to avoid the conclusion that more effective action is achieved where there is more effective collaboration between the federal and provincial or state governments, for whatever reasons.

None of the above discussion is intended to imply that, in either the Great Lakes basin ecosystem or the Georgia Basin–Puget Sound bioregion, overall progress in addressing environmental issues of concern is as extensive as it could or should have been. In the former, formalized institutional arrangements linking the two federal governments and others linking the provinces/states and their respective federal governments have been helpful in effecting cooperation. In the latter, federal-federal institutional arrangements have found only limited use. Rather, there has been an emphasis on arrangements that involve major province or state participation and often leadership. These arrangements can be with the respective federal governments, as in the case of the Puget Sound Water Quality Authority or the Fraser Basin Management Board (both of which also involve municipal and First Nations participation), or they may be direct province- or state-led approaches, as discussed below. The salient lesson is simply that, when there is collaborative effort, a greater degree of progress is possible, regardless of formal institutional arrangements.

State-Provincial Relationships

In both the Great Lakes basin ecosystem and the Georgia Basin–Puget Sound bioregion, formal institutional arrangements have been established at the state-provincial level (see governance flowcharts). In the Great Lakes region, the Interstate Council of Great Lakes Governors includes the premier of Ontario as an observer. In addition, Ontario and the eight Great Lakes states have signed both the Great Lakes Charter and the Toxic Substance Control Agreement. In 1992, British Columbia and Washington signed an umbrella agreement on cooperation, dealing mainly with economic development. Jamie Alley's discussion elsewhere in this volume provides a detailed description of the evolution of ECC, also created in 1992. British Columbia and Washington further signed a series of agreements in 1994 to facilitate joint action on transportation, growth management, and technology- and knowledge-based industries.

By the spring of 1997, implementation of commitments emerging from the work of the ECC had slowed due to government cutbacks, particularly in British Columbia, although the task forces on marine reserves, toxic contaminants, and monitoring of the aquatic ecosystem continued to meet. The June, 1997, collapse of the U.S.-Canada negotiations on the Pacific Salmon Treaty led the government of British Columbia to suspend its crossborder activities. As a result, what little momentum the ECC had achieved came to a halt. However, the individuals who created the ECC are still in place, as is their commitment to cooperative activity to solve common problems. In short, whether or not the ECC emerges from its current state of suspension, a cooperative foundation is in place that will shape future activities.

Because the ECC is a body created by province and state, federal representatives sit at its meetings as observers. However, federal agencies (in particular, the U.S. Environmental Protection Agency, Environment Canada, and Fisheries and Oceans Canada) have played major roles in program implementation. Even in the absence of British Columbia's participation in discussion of issues of concern to both British Columbia and Washington, federal representatives have continued the crossborder dialogue.

Local-Local Government Relationships

In the Great Lakes, municipal governments are linked formally through the International Great Lakes–St. Lawrence Mayors Conference. No equivalent exists in the Georgia Basin–Puget Sound bioregion, although two local transboundary linkages recently have emerged. In October, 1995, the first meeting of the Alliance of Border Communities brought together a coalition of communities of British Columbia's Fraser Valley and Washington State's Whatcom and Skagit counties to pool information and exchange ideas on mutual problems. Delegates included mayors, business leaders, representatives of each of the three community colleges, environmentalists, and arts leaders. Three river valleys make up the region: the Fraser, the Skagit, and the Nooksack. The organizing agency is the Columbia Pacific Foundation.[50] Exploratory meetings also have been held between the Islands Trust, which has jurisdiction over the Gulf Islands, and counties in the San Juan Islands on the U.S. side. Cooperative activity on technical matters occurs informally between adjacent local governments along the border as need arises.

The Role of First Nations

In both the Georgia Basin–Puget Sound bioregion and the Great Lakes basin ecosystem, First Nations communities span the international boundary. An example is the Fort William Band in Thunder Bay and its links to the Pigeon Band in northern Minnesota.

While much current effort is focusing on negotiations of land claims treaties, particularly in Canada, the issue of water resources and related First Nations rights rarely has been addressed, let alone resolved. Further, the role of First Nations people in comanaging resources is rapidly growing in significance. Because traditional native territories bear little resemblance to the current location of the U.S.-Canada border, and because transboundary relationships are maintained to this day, First Nations' potential for serving as a catalyst for enhancing transboundary cooperation is great.

BUILDING COMMUNITY

Environmental Nongovernment Organizations (ENGOs)
Great Lakes United (GLU), a formal, binational citizens' coalition, has grown since its creation in 1982 to include over two hundred member groups of environmentalists, sports enthusiasts, labor, and civic organizations from the eight Great Lakes states, Ontario, and Québec. GLU's emergence was an important step forward in dealing with Great Lakes environmental issues. GLU has offices in Buffalo and Montréal and has served to focus and enhance public interest and effort through a broad range of programs aimed at conserving and protecting the Great Lakes–St. Lawrence River ecosystem.

In addition to GLU and its members, dozens of unaffiliated, locally based organizations are scattered around the lakes. In addition, both Canadian and American national and provincial or state organizations have a regional focus on the Great Lakes (for example, the Conservation Council of Ontario, Audubon Society, Greenpeace, Sierra Club, Nature Conservancy, Friends of the Earth). A large number of subregional, community-based roundtables (formal and informal gatherings of people from across communities aimed at collaborative solution-building) also serve to draw together those with like interests. The overall result is an extremely rich and active network of ENGOs.

In the Georgia Basin–Puget Sound bioregion, the development of a formal coalition of ENGOs began in the early 1990s, with the creation of the Save the Georgia Strait Alliance (later renamed the Georgia Strait Alliance) on the Canadian side and People for Puget Sound on the U.S. side. In 1995, the two joined together to establish the Sound and Straits Coalition. The coalition links more than fifty groups from around the bioregion. However, it has not yet developed the level of human and financial resources or the influence enjoyed by Great Lakes United.

As in the Great Lakes basin ecosystem, a number of subregional, community-based roundtables are active in the bioregion. However, the overall web of ENGOs in the Georgia Basin–Puget Sound bioregion is in an earlier phase of development than in the Great Lakes.

Transboundary Collaborative Research

In the Great Lakes, transboundary collaboration among researchers and academics has been an important source of coalition building. Much of this has been facilitated through the boards and task forces of the IJC and the GLFC. In addition, the International Association of Great Lakes Research (IAGLR) meets annually and publishes a peer-reviewed technical journal. A number of universities have institutes focusing on Great Lakes issues.

There is no equivalent formalized network of collaborative research in the Georgia Basin–Puget Sound, although informal linkages do exist and are manifested in joint projects and conferences. One of the ECC's task forces has been charged with developing a coordinated transboundary research and monitoring capacity. No program has yet resulted.

In both the Great Lakes area and the Georgia Basin–Puget Sound bioregion, major concern has been voiced about the erosion of support for research. An IJC report shows that the total budget and number of researchers from thirty-one institutions around the Great Lakes, representing 80 percent of the total funding in the inventory of the IJC's Council of Great Lakes Research Managers, have been dropping and are projected to continue to drop.[51]

The IJC has pointed out that these kinds of cuts could seriously compromise the governments' ability to achieve commitments made under the Great Lakes Water Quality Agreement.[52] Although similar figures for the Georgia Basin–Puget Sound have not been compiled, anecdotal evidence suggests that the same sort of erosion of support is occurring there. For example, the Defense Research Establishment (Pacific) in Victoria was closed in 1994, as was the West Vancouver laboratory of Fisheries and Oceans Canada. The Pacific-Yukon Region of Environment Canada experienced a 25-percent budget cut in fiscal 1995–96; several divisions of Fisheries and Oceans Canada experienced greater than 50-percent cuts between 1992 and 1997.

These losses seriously undermine the basis for effective decision making. Loss of scientific memory and gaps in monitoring threaten future trends analysis of ecosystem integrity. Moreover, reductions in the research community are contributing to an erosion of transboundary collaboration and its attendant sense of community. Such changes may have the most detrimental implications of all in terms of achieving constructive change.

The Sense of Community

A remarkable characteristic of the Great Lakes basin ecosystem is the sense of identity that residents feel with the Great Lakes. In consequence, a regional sense of community exists that transcends political borders. The rich networks of community organizations; the professional relationships that have been developed through the IJC, GLFC, government, and academic channels; and the many resultant publications all have contributed to the emergence of a "Great Lakes Community." There is no doubt that the biennial meetings of the IJC have served as a matrix for the formation of this community, as well as providing an important mechanism for two-way communication between stakeholders and government. In turn, it is within the community that the impetus for change is seeded and nourished.

The beginnings of such a community now are evident in the Georgia Basin–Puget Sound bioregion, spurred on by a number of factors:

- The continuing work of the Puget Sound Water Quality Authority, including sponsorship of various meetings, such as Puget Sound Forum 1995, in which over seven hundred citizens, scientists, and environmentalists participated.
- Creation of the ECC in 1992 and the establishment of a number of subsidiary groups, including the Marine Sciences Panel.
- Continuing work by the Pacific Northwest Economic Region (PNWER), which focuses on economic development but also has dealt with recycling.
- A number of transboundary conferences and symposia, including the Symposium on the Marine Environment (under the auspices of the ECC's Marine Sciences Panel), held on January 13–14, 1994.
- Creation by the British Columbia government of the Georgia Basin initiative in 1993, following on the recommendation of the now-defunct British Columbia Roundtable on the Environment and the Economy.
- Continuing efforts of the Canadian Studies Program at the University of Western Washington, Bellingham, to effect collaboration through U.S.-Canada (and often Mexico) conferences, workshops, and exchanges.
- Emergence of the Sound and Straits Coalition.
- Informal meetings between representatives of the Islands Trust on the British Columbia side and the San Juan Islands on the Washington side, to discuss a range of issues of concern.
- Development of the Pacific Marine Heritage Legacy, a five-year federal provincial program for creating an expanded and integrated system of coastal marine parks.
- Establishment in 1990 of the Pacific Coast Joint Venture, to coordinate the efforts of government agencies and private organizations trying to protect and manage Pacific Coast wetlands and adjacent upland habitats.

However, a number of significant differences exist between the Great Lakes and the western bioregion that will affect development of a sense of community. First, the bioregion does not enjoy the same level of symbolic prominence that the Great Lakes as a region do. The Great Lakes have been recognized and photographed from outer space; they have been described as an important part of the global heritage and as one of the "absolute values our planet possesses."[53] Among residents of the Great Lakes basin ecosystem, there is a deeply entrenched sense that they live in the heartland of North America, and they identify with this heartland.

In contrast, within the Georgia Basin–Puget Sound bioregion, the sense of place is fragmented. Residents of Vancouver Island, Howe Sound, the lower mainland and Vancouver, the Gulf Islands, the San Juan Islands, Puget Sound, or Seattle-Tacoma all identify with their particular locality rather than with the bioregion as a whole. In general, they are part of the "West Coast," and there is a palpable "West Coast ethos." Cut off by the mountains from the rest of North America, their perspective long has been toward the Pacific Ocean, despite eastward railway, road, and air connections. The market for both Boeing and Microsoft, located close to Seattle, is the world, not the continent. Efforts that focus on the whole of the "Northwest" or "Cascadia" as a development area, rightly or wrongly, divert attention from the Georgia Basin–Puget Sound bioregion.[54]

Further, the bioregion lacks organizations equivalent to the IJC and GLFC, with their boards, task forces, and meetings. No body like the International Association of Great Lakes Research, with its publications program, spawns transboundary collaborative research. Similar work by the ECC is still in an embryonic stage. On the Canadian side, no federal-provincial agreement exists to facilitate collaboration. At the local level, there is nothing like the International Great Lakes–St. Lawrence Mayors' Conference to facilitate discussion of municipal concerns throughout the region.

The power of a sense of community is very clear. People react to threats to their community and are motivated to act as a result of concern for, and pride in, their community. There are direct links among the sense of community, public policy formation, and subsequent action. When concern and pride strengthens and persists, politicians respond.

CONCLUSIONS: SUSTAINABILITY AND LINKING HUMAN AND ECOSYSTEM WELL-BEING

Progress toward sustainability builds from the goals of maintaining both human and ecosystem well-being and improving it. The idea of sustainability implies that certain features of the world need to be retained, if life (for people, plants, and animals) is not increasingly to be degraded. This idea is an expression of interdependence, and it draws on values that differ from place to place and change over time. Because of this dependency upon values, there is no unique design for a sustainable world.

These thoughts are not complicated. However, they do commit society to considering the long term, to recognizing peoples' places within the ecosystem, and to reflecting consciously upon implications of human activity for the well-being of both people and ecosystems. Together these ideas provide a new lens for seeing the world, a lens that forces us to integrate many ideas and disciplines that previously have remained disparate. Those using this lens, including the World Commission on Environment and Development, have come to the conclusion that the current nature of human activity is seriously undermining opportunities for future generations.[55] That is why they called for a new way of doing things, a way that was forward-thinking, to insure that what people did today did not undermine opportunities for generations not yet born.

In this brief comparative review, the lens of sustainability has shaped the perspective of our analysis. The following conclusions emerge. In terms of traditionally defined economic success, as measured by per-capita productivity, the Great Lakes basin ecosystem and the Georgia Basin–Puget Sound bioregion now are essentially the same. Only recently, however, has attention turned to accounting for the full human and ecosystem costs of that success. From such an overall perspective, the Great Lakes basin ecosystem may be at a critical point in its evolution. Past growth and economic expansion may be coming to an end, to be replaced by either a stable population and a more steady-state economy, or a period of decline. Ecosystem restoration is emerging as a key to economic and social transformation. It is worrisome, to say the least, that reductions in government support for Great Lakes–related research and action come at this critical transition point. Will the progress of the past twenty-five years be replaced by institutional apathy and a falling away from the ambitious, even noble, goals of the ecosystem management approach that was enshrined in the 1978 Great Lakes Water Quality Agreement? Can the will to act be sustained under such conditions?

Much of the early success in environmental restoration in the Great Lakes saw significant gains for relatively small expenditures of resources. This has been the easy phase. Many of the environmental issues that remain to be addressed—for example, the lingering levels of persistent toxic contaminants—are tougher and more expensive to deal with, but no less important. More than anywhere else in the world, perhaps, the management regime of the Great Lakes has moved beyond reactive crisis management to a more anticipatory, forward-looking regime. Will this capacity now atrophy, just as community support and organizational capacities have evolved to facilitate dealing with the more difficult and complex phases of resolving environmental issues of concern?

In the Georgia Basin–Puget Sound bioregion, economic and social expansion, much of it driven by Pacific Rim and other international trade, is continuing at high rates. Here the challenge is to continue on this path without degrading the livability and natural environment of the region, as was done in the expansionist period of Great Lakes development. In both Washington and British Columbia, growth management acts have been established that have no counterparts in the Great Lakes basin. In Washington, this has led to strict delineation of areas available for urban expansion and areas to be maintained as rural. In British Columbia, establishment of the Agriculture Land Reserve (ALR) over twenty years ago has had a significant impact in limiting urban sprawl. Recently the intent of the ALR has been buttressed by the Farm Practices (Right to Farm) Protection Act, as well as the establishment of the Forest Land Reserve, although a 1998 exclusion in the interior of the province raised a significant debate about the ALR's future. In any case, the success and implications of these initiatives remain to be discovered.

Both regions are facing many parallel concerns regarding environmental degradation and opportunities for restoration. Examples include: air quality, surface water quality, persistent toxic substances, the current crisis in the fishing industry, and habitat and agricultural land destruction related to urban sprawl. Also in both regions, some successes now are evident in dealing with these environmental issues, particularly the reduction of certain emissions and discharges and a resulting improvement in ecosystem conditions. However, in both regions many of the physical and biological changes that have been imposed by human activity remain to be reconciled with ecosystem integrity. Overall, not even the maintenance of human and ecological well-being is assured, let alone its improvement.

In terms of achieving action, it is clear that the political will to act derives, as much as anything, from a sense of place and community and attendant pressures being brought to bear on the public policy process. Public policy aimed at cultivating this sense of place and community will facilitate action sooner rather than later.

Finally, collaborative federal-provincial/state approaches for dealing with transboundary environmental concerns seem to be more effective than independent activities by either of the parties. In some cases, informal working relationships are the key to success. However, when many elements of government and civil society are implicated, as is usually the case in addressing transboundary environmental issues, overall coordination and collaboration to develop solutions that avoid duplication and waste are needed. These ends are most likely to be achieved with a more formal institutional arrangement. Now more than ever, with financial resources at all levels of government increasingly restricted, successful collaboration is required.

In the Great Lakes region, this conclusion would argue for a strengthening of the Great Lakes Office of the IJC. Currently the opposite appears to be taking place. And in the Georgia Basin–Puget Sound Bioregion, a strengthening of the ECC, rather than the weakening that has occurred, would seem warranted. On the Canadian side of the bioregion, federal-provincial collaboration and coordination need to be enhanced. For all of these matters, the required direction of action seems clear, but the political will to act is less evident.

NOTES

1. Robert A. (Tony) Hodge, "Assessing Progress Toward Sustainability: Development of a Systemic Framework and Reporting Structure," Ph.D. diss., McGill Univ., Montréal, Canada, 1995, p. 173.
2. World Bank, *World Development Report, 1992: Development and the Environment* (New York: Oxford Univ. Press, 1992), 222.
3. Summarized from Hodge, "Assessing Progress Toward Sustainability," ch. 9.
4. Environment Canada, EPA, Brock Univ., and Northwestern Univ., *The Great Lakes: An Environmental Atlas and Resource Book* (Chicago and Toronto: U.S. EPA and Environment Canada, 1988), 17.
5. Phil Weller, *Fresh Water Seas: Saving the Great Lakes* (Toronto: Between the Lines, 1990), 41.
6. D. Braider, *The Niagara* (New York: Holt, Rinehart and Winston, 1972).
7. William A. Testa, "Overview," in "The Great Lakes Economy—Looking North and South," pp. v–xi in Federal Reserve Bank of Chicago and the Great Lakes Commission, 1991 (Chicago: Federal Reserve Bank of Chicago, 1991).
8. John Ryan, "State of the Northwest," Northwest Environmental Watch Report no. 1 (Seattle: Northwest Environment Watch, 1994), 15.
9. Gross domestic product figures for Washington State taken from U.S. Bureau of Census, *Statistical Abstract of the U.S., 1995* (Washington, D.C.: Commerce Department, Census Bureau, 1995). Those for British Columbia are from British Columbia Ministry of Finance and Corporate Relations, *1995 British Columbia Financial and Economic Review* (Victoria: Province of British Columbia, 1995). Conversion from Canadian to U.S. dollars, using the average exchange rate for 1990 of .8571, taken from Statistics Canada.

10. These population figures are for the entire eight Great Lakes states and Ontario, not just their in-basin portions.
11. U.S. Bureau of the Census, *Statistical Abstract of the United States, 1995* (Washington, D.C.: Department of Commerce, 1995); B.C. Ministry of Finance and Corporate Relations, *1995 British Columbia Financial and Economic Review* (Victoria: Province of British Columbia, 1995). Statistics Canada, National Income and Expenditure Accounts. Quarterly Estimates, First Quarter 1992. Catalog 13-001 Quarterly (Ottawa: Minister of Industry, Science and Technology, 1992).
12. British Columbia Roundtable on the Environment and the Economy, "Georgia Basin Initiative: Creating a Sustainable Future" (Victoria: British Columbia Roundtable on the Environment and the Economy, 1993), 6.
13. Bruce Agnew, "Overview of Washington State and Perspective on Cross-Border Issues" (Victoria: British Columbia Roundtable on the Environment and the Economy, 1992).
14. Ryan, "State of the Northwest," 5.
15. International Joint Commission, Docket 4R, *Pollution of Boundary Waters* (Washington, D.C., and Ottawa, 1912).
16. International Joint Commission, Docket 54R, *Pollution of St. Clair River, Lake St. Clair and Detroit River and St. Mary's River,* 1946; Docket 55R, *Pollution of Niagara River,* 1948; Docket 61R, *Air Pollution in Windsor-Detroit Area from Vessels,* 1949; Docket 83R, *Pollution of Lower Great Lakes,* 1964; Docket 85R, *Air Pollution,* 1966 (Washington, D.C., and Ottawa).
17. Theo Colborn, Al Davidson, Sharon Green, Tony Hodge, Ian Jackson, and Rich Liroff, *Great Lakes, Great Legacy?* (Washington, D.C.: Conservation Foundation; and Ottawa: Institute for Research on Public Policy, 1990), xix.
18. Environment Canada, "Great Lakes: Pulling Back from the Brink," ch. 18 in *The State of Canada's Environment* (Ottawa: Minister of Supply and Services Canada, 1991).
19. Hodge, "Assessing Progress Toward Sustainability."
20. Environment Canada, "Great Lakes: Pulling Back from the Brink," 18–20.
21. IJC, *8th Biennial Report on Great Lakes Water Quality* (Ottawa and Washington, D.C.: IJC, 1996), 2–3.
22. B.C.-Washington Marine Science Panel, *The Shared Marine Waters of British Columbia and Washington* (Victoria: Government of British Columbia; and Olympia: State of Washington, 1994). See also Bob Wilson, R. J. Beamish, F. Aitkens, and J. Bell, "Review of the Marine Environment and Biota of Strait of Georgia, Puget Sound and Juan de Fuca Strait," in *Canadian Technical Report of Fisheries and Aquatic Sciences, 1988* (1994), 398.
23. Puget Sound Water Quality Authority, *State of the Sound, 1986 Report* (Seattle: Puget Sound Water Authority, 1996); see also *1987 Puget Sound Water Quality Management Plan* and the *1991 Puget Sound Water Quality Management Plan* (Seattle: Puget Sound Water Authority).
24. State of Washington, Department of Ecology, and EPA, *Environment 2010:*

The State of the Environment Report (Olympia: State of Washington, 1989).
25. Province of British Columbia and Environment Canada, *State of the Environment Report for British Columbia* (Victoria: B.C. Ministry of Environment, Lands, and Parks and Parks and Environment Canada, 1993).
26. British Columbia Roundtable on the Environment and the Economy, *State of Sustainability: Urban Sustainability and Containment* (Victoria: Crown Publications, 1994).
27. British Columbia Ministry of Environment, Lands, and Parks, "Report on British Columbia Progress Toward Sustainability" (Victoria: B.C. Ministry of Environment, Lands, and Parks, 1997).
28. Fraser Basin Management Program, Board Report Card 1996 (Vancouver: Fraser Basin Management Program, 1996).
29. Ryan, "State of the Northwest."
30. Theo Colborn et al., *Great Lakes, Great Legacy?* 122.
31. Province of British Columbia and Environment Canada, *State of the Environment Report for British Columbia* (Victoria: 1993), 37.
32. Puget Sound Water Quality Authority, *1991 Puget Sound Water Quality Management Plan* (Olympia, Wash.: Puget Sound Water Quality Authority, 1990), 21.
33. Theo Colborn, D. Dumanoski, and J. P. Myers, *Our Stolen Future* (New York: Dutton Books, 1996); Theo Colborn and C. Clement, eds., *Chemically Induced Alterations in Sexual and Functional Development: The Wildlife-Human Connection* (Princeton, N.J.: Princeton Scientific Publishing Co., 1992). IJC, *Sixth Biennial Report on Great Lakes Water Quality, 1992; Seventh Biennial Report on Great Lakes Water Quality, 1994;* and *Eighth Biennial Report on Great Lakes Water Quality, 1996* (Washington and Ottawa: IJC).
34. IJC, *8th Biennial Report,* 8–10.
35. Figures for Puget Sound are given in: Puget Sound Water Quality Authority, *State of the Sound 1986 Report* (Olympia, Wash.: Puget Sound Water Quality Authority, 1986), 32. Figures for the Great Lakes are summarized in Theo Colborn et al., *Great Lakes, Great Legacy?* 143–44.
36. Royal Commission on the Future of the Toronto Waterfront, *Regeneration: Toronto's Waterfront and the Sustainable City: Final Report* (Toronto: Royal Commission on the Future of the Toronto Waterfront, 1991), 115.
37. Royal Commission on the Future of the Toronto Waterfront, *Environment in Transition,* Discussion Paper no. 10 (Toronto: Royal Commission on the Future of the Toronto Waterfront, 1990), 113.
38. Hodge, "Assessing Progress Toward Sustainability," 288; G. R. Francis, "Binational Cooperation for Great Lakes Water Quality: A Framework for the Groundwater Connection," *Kent Law Review* 65, no. 2 (1989): 359–73.
39. Great Lakes Fishery Commission, *Annual Report* (Ann Arbor, Mich.: Great Lakes Fishery Commission, 1983).
40. Carlos Fetterolf, Jr., *A Sketch of the Great Lakes Fishery Commission: Interagency, Interstate, International Programs* (Ann Arbor, Mich.: Great

Lakes Fishery Commission, 1986). See also Colborn et al., *Great Lakes, Great Legacy?*, 148–63.

41. Pacific Salmon Treaty (1985). See also Carl Walters, *Fish on the Line: The Future of Pacific Fisheries* (Vancouver: David Suzuki Foundation, 1995), 18.
42. Boundary Waters Treaty of 1909, Article IV.
43. Lynton K. Caldwell, ed., *Perspectives on Ecosystem Management for the Great Lakes* (Albany: State Univ. of New York Press, 1988).
44. Rich Thomas, Jack Vallentyne, Ken Ogilvie, and J. D. Kingham, "The Ecosystems Approach: A Strategy for the Management of Renewable Resources in the Great Lakes Basin," ch. 1 in Caldwell, *Perspectives on Ecosystem Management for the Great Lakes*, 33.
45. E.g., see Environmental Law Institute, "An Evaluation of the Effectiveness of the International Joint Commission" (Washington, D.C.: Environmental Law Institute, 1995).
46. The ideas in this section are taken mainly from Colborn et al., *Great Lakes, Great Legacy?* 194–97.
47. Auditor General of British Columbia and Deputy Minister's Council, *Enhancing Accountability for Performance in the British Columbia Public Sector* (Victoria: Office of the Auditor General, 1995); Auditor General of British Columbia and Deputy Minister's Council, *Enhancing Accountability for Performance: A Framework and Implementation Plan: 2nd Joint Report* (Victoria: Office of the Auditor General, 1996).
48. See Mel Smith, *The Renewal of Federation: A British Columbia Perspective* (Victoria: Ministry of Provincial Secretary, 1991).
49. Puget Sound Water Quality Authority, "Issue Paper: Public Involvement in Water Quality Policy Making—June 1986" (Seattle: Puget Sound Water Quality Authority, 1986), 1.
50. Fax from Joy Monjure, Columbia Pacific Foundation, Bellingham, Wash., to John Wirth, Aug., 1996.
51. IJC, *8th Biennial Report*, 5–7.
52. Ibid., 5.
53. N. N. Moiseev, "Statement to the World Commission on Environment and Development Public Hearing, Dec. 8, 1986, Moscow, Russia," noted on page 285 of World Commission on Environment and Development, *Our Common Future* (New York: Oxford Univ. Press, 1987), 285.
54. For example, the work on the Cascadia Concept championed by the Cascadia Institute. The institute includes a cadre of intellectuals drawn from both Washington State and British Columbia.
55. World Commission on Environment and Development, *Our Common Future*.

4

GREAT WHALE
FROM CONFLICT TO JOINT PLANNING OF QUÉBEC'S ENERGY POLICY

DAVID CLICHE, WITH LUCIE DUMAS

Before addressing the subject of this chapter, a brief discussion of some of the highlights of Québec's relations with the United States in the areas of energy and the environment is in order. As early as July, 1982, Québec and the State of New York signed an agreement on acid rain, marking the first bilateral agreement of this kind between a Canadian province and an American state. Subsequently, in 1985, the State of New York implemented a program to reduce sulphur dioxide emissions by 1998. The Great Lakes Charter, signed in 1985, made Québec, Ontario, and eight waterfront states partners in improving water quality management of the Great Lakes. Management of the St. Lawrence River and lakes Memphrémagog and Champlain, all transboundary bodies of water, also has been the subject of much discussion and negotiation between the bordering Canadian provinces and American states. Québec and Vermont, for example, are parties to a cooperation agreement for management of Lake Memphrémagog, while Québec, Vermont, and New York are signatories of an agreement on the protection of Lake Champlain.

To the people and the government of Québec, environmental protection is more than just a passing trend; it is a crucial responsibility that Québec assumes in numerous ways, both within its own borders and through international endeavors. Given the scope of the energy and environmental issues of concern to Québec and the northeastern states, it is my opinion that existing forums provide perfect opportunities to address these issues in greater detail.

ELECTRICITY WITHOUT BOUNDARIES

The export of electricity from Canada has been the topic of endless debate over the past thirty years. Great Whale was the culmination of the controversy surrounding electrical power generation, not only in

Québec and the rest of Canada, but also in the U.S. and Europe. How the six-year debate (1989–95) over this hydroelectric project changed the decision-making process for energy projects in Québec and led to greater involvement of interested groups is what we are about to see.

LAUNCHING THE GREAT WHALE PROJECT: FUTILE DEBATE

Québec's energy policy under the Liberal government of Robert Bourassa was aimed specifically at providing massive electricity exports to the U.S., based on firm energy contracts spanning several years. In May, 1989, the government launched the Great Whale project (James Bay 2), in the hope of exporting a minimum of ten thousand megawatts of firm energy south of the border, depending on demand. Ominously, the minister of energy and resources warned Québec's citizens that, without Great Whale, they never would have enough energy and that, if the project did not go ahead, they eventually would have to light their houses by candlelight. Hydroelectricity was the sole form of energy promoted by the Bourassa government. The "great power of the North" dictated the government's 1988 energy policy, the explicit goal of which was to export ten thousand megawatts of electricity to the U.S.

By fall, 1990, however, it had become clear that Québec's export-oriented energy policy no longer was attuned to market conditions in the northeastern U.S. Instead, it was the dream of politicians who refused to admit that there was an energy glut. Since the introduction of the Public Utilities Regulatory Policies Act of 1978 (PURPA), independent power producers had been taking over a growing share of the production market. In addition, the Energy Policy Act of 1992 gave independent producers access to power transmission lines owned by the public utilities.

In short, the U.S. electrical power industry was undergoing a veritable revolution, marked by fierce competition among producers and distributors, the emergence of small producers and distributors, and the cancellation of megaprojects. Long-term purchasing contracts were a thing of the past. However, Hydro-Québec rapidly adjusted to this new market. Total electricity sales to the U.S. grew, and seasonal sales increased proportionately.[1]

THE BATTLE LED BY QUÉBEC CREES

Few analysts in Québec took time to explain the new energy context south of the border. As a result, cancellation of long-term contracts between Hydro-Québec and the New York Power Authority (NYPA) generally was attributed to protests led by James Bay Crees, rather than to existing market forces. The Québec Crees achieved near-instant fame in the northeastern U.S., becoming the focal point of a complex energy issue. However, a close reading of the documents pertaining to the NYPA's decision to cancel its contracts with the Québec utility confirms that the decision was based primarily on economic reasons, although environmental protection and the Native peoples also are mentioned.

On March 27, 1992, NYPA's Chair Richard M. Flynn announced that he had terminated a contract for the purchase of one million kilowatts of hydroelectric power from Hydro-Québec. He said that economic considerations led to cancellation of the agreement signed in 1989. The press release added: "By last August, when Flynn negotiated an 11-month delay in the deadline for a final decision on the contract, it was clear the economics had changed significantly. Key factors in the change, Mr. Flynn said, were sizable increases in the projected roles of energy conservation and independent power production and a sharp decrease in forecasted prices for oil and natural gas."[2]

The Great Whale project, located on the shores of Hudson Bay 150 kilometers north of the La Grande complex, entails the construction of three electric power plants on the Great Whale River and the diversion of the Little Whale River. A project of this scope also requires construction of permanent roads and several housing units, as well as an entire system of power transmission lines. The reservoirs alone would have covered an area of approximately four thousand square kilometers. The generating capacity of the complex was estimated at 3,100 megawatts and the cost of construction at sixteen billion dollars (in 1991 Canadian currency).

The Crees opposed the Great Whale project for two fundamental reasons. The Great Whale project came at a time when the Québec Crees felt they no longer could assimilate the social upheaval caused by hydroelectric development. In the space of thirty years, they had gone from a nomadic lifestyle to a sedentary one. Having been a forgotten nation in northern Québec, in the 1970s they, along with the Inuit, suddenly became the focus of Québec's attention during the negotiation of the James Bay and Northern Québec Agreement (JBNQA). When Great Whale came along some fifteen years later, the Crees viewed it as a symbol of social problems that could only get worse. They felt they had no choice but to try to delay or even block the project, in order to slow down or stop the process of radical social change that was affecting so many aspects of their lives.

A second reason for their fierce opposition to Great Whale was rooted in the fact that the Crees had hired consultants who lost no time in jumping on the political and economic bandwagon. These consultants initiated what would be for them an extremely lucrative international campaign against the Great Whale project; at its height, the protest campaign cost the Crees some two million dollars a year.

The Québec government at the time blamed the Crees for the cancellation of Hydro-Québec's contracts with the American states. The Crees then broke off relations with the government of Québec. Moreover, a number of U.S. senators and government representatives had championed the cause of the Native peoples and the rivers of northern Québec. In reality, however, they were protecting U.S. power producers. Held in high esteem by some and in contempt by others, the chief of the Grand Council of the Crees of Québec, Matthew Coon Come, and Robert Kennedy, son of the late Bobby Kennedy, became the torchbearers for a cause whose true protagonists were unknown to the general public. In a cunning move, Canadian producers and exporters of natural gas remained invisible to the media during the Great Whale debate, taking advantage of the situation to sign numerous energy supply contracts with independent American producers, using gas turbines. Opposition to the Great Whale project coincided with the opening up of the U.S. electricity market and the emergence of independent producers.[3]

THE GOVERNMENT OF CANADA

As soon as the project was launched in 1989, the government of Canada announced that it would be conducting its own environmental assessment of Great Whale, since the project was expected to impact on areas under federal jurisdiction, such as marine mammals and migratory birds, and to have consequences beyond Québec's borders, since nearby rivers flow into Hudson Bay, which is under federal jurisdiction.

The federal environment minister at the time, Lucien Bouchard, and his negotiator, David Cliche (the author of this chapter), explained that Ottawa was obliged by the courts to enforce its own environmental impact review processes. (A prime example of the complexity of the Canadian federal government is the fact that Lucien Bouchard now is the premier of Québec, while your author now serves as Québec's Minister of Tourism and has served as Minister of Environment and Wildlife, whose job is to plead the case of Québec jurisdiction to Ottawa.)[4] Québec unwillingly accepted the situation; and, in the fall of 1989, negotiations began in a bid to harmonize the five different environmental impact assessment and review procedures then in effect.

OTHER PLAYERS

Workers belonging to the Fédération des travailleurs québécois (Union of Québec Workers) and the Canadian Manufacturers' Association (Québec branch) formed a coalition in favor of Great Whale, while environmental groups, such as Greenpeace, called for a moratorium on the project and a public energy debate. Opponents of the project saw Great Whale as unnecessary and as a threat to the idealized way of life of the Native people, while its supporters saw the project as essential to Québec's economic development.

In fact, Greenpeace Québec and other Québec groups had established links with U.S. groups, especially the National Resources Defense Council. These U.S. environmental groups (NRDC, Audubon Society, Conservation Law Foundation) strongly supported the Crees' cause from south of the border. And when, in the fall of 1989, the Québec Crees made the front page of the *New York Times,* the stage was set for major conflict. It was not long in coming.

GREAT WHALE FORUM

The Great Whale Forum was born of my exasperation with the massive protest on the one hand and assertions of the urgent need to go ahead with the project on the other. I felt that the situation was getting out of hand and that the high-decibel debate was unproductive. Surely there had to be a way to discuss the Great Whale project calmly, without the counterproductive antagonism which had characterized the whole debate up to that time.

As president of a firm specializing in the environment and in integrated resource planning, I had fifteen years experience consulting on issues pertaining to the James Bay territory and northern Québec in general. In January, 1991, I decided to found the Forum québécois pour l'examen public du complexe Grande-Baleine (Québec Forum for the Public Examination of the Great Whale Complex) and presided over it until my election to the National Assembly of Québec in 1994.[5]

At the time, the Québec government was in a hurry to begin construction of Great Whale and had decided to divide the environmental review process into two parts. The access roads and auxiliary infrastructures would be reviewed first, followed by the dams and reservoirs. The camp in favor of the project maintained that delaying construction would put the entire project at risk. Project opponents, on the other hand, predicted disastrous consequences if construction were to go ahead before an environmental review had been completed. Until then, the Great Whale project had been discussed at length both within and outside Québec, without ever touching on the fundamental aspects of the project, such as the need for it, its impacts both positive and negative, and the alternatives. I was hoping that the Great Whale Forum would bring together organizations fighting for a stringent public review of the proposed hydroelectric project. The forum was to be an umbrella organization representing the following non-partisan groups: Laval University's Centre d'études Inuit et circumpolaires (Inuit and Circumpolar Studies Center), the Union québécoise pour la conservation de la nature (UQCN), the Fédération québécoise de la faune (FQF) (Québec Wildlife Federation), the Assemblée des évêques du Québec (Assembly of Québec Bishops), the Association québécoise des biologistes, the Coopératives d'économie familiale, the Québec branch of the Confederation of National Trade Unions (CSN) and the Union des producteurs agricoles (UPA).

The Great Whale Forum played a key role in sparking public debate on the fundamental aspects of this hydroelectric project and in obtaining a guarantee that the project would undergo a comprehensive environmental review. Thanks to this process, a number of precedents were set in Québec. Initially, the Québec government opposed this forum's creation; however, eventually, albeit very reluctantly, it recognized the usefulness of such a body. The roles played by the Québec Crees and a few other groups also helped achieve these objectives.

The forum held four conferences on the main aspects of the project's environmental review. At the first conference, which took place in June, 1991, we defined the environmental review procedure standards that forum members felt should be applied in carrying out the public review of Great Whale. Then we publicly lobbied the governments of Québec and Canada and the other parties concerned to get them to enforce these environmental review standards recommended by the forum. They did this.

At the second forum conference, in December, 1991, the human and biophysical environment of the Great Whale River was discussed. Over a two-day period, we brought together representatives of the Crees, the Inuit, Hydro-Québec, and a number of federal and Québec government departments, as well as university researchers, to share their knowledge and expertise. For the first time in years, all of the parties felt free to exchange viewpoints in an atmosphere of cooperation, without confrontation.

Four months later, in April, 1992, the forum organized a third conference on the planning and justification of hydroelectric projects. Specialists from around the globe, including scientists from the U.S., were invited to participate in the discussions. The aim of this conference was to identify, evaluate, and integrate the various economic, social, and environmental criteria for hydroelectric development in Québec. This was the first conference on integrated resource planning (IRP) ever held in Québec. Although the approach already was being used widely by public utilities in the U.S., it was new to Québec. The conference therefore served to introduce two new methodologies: collaborative processes and IRP.

The theme of the fourth and last conference, held in November, 1992, was cumulative environmental and social impact assessment methods. Experts from Québec, Canada, and the U.S. shared their knowledge and compared their experiences in order to determine the type of impact assessment that should be applied to hydroelectric projects in the James Bay and Hudson Bay regions. Representatives from Hydro-Québec, Ontario Hydro, Manitoba Hydro, and the Canadian Arctic Resources Committee explained their research programs on cumulative impact assessment of development in these two regions. The upshot was that everyone involved expressed satisfaction with the leadership shown by the forum in the public debate over Great Whale, one of the most extensive and significant debates of its kind to take place in Québec in the last decade. All groups contributed financially to the forum's activities—that is, the Crees, the Inuit, the governments of Québec and Canada, Hydro-Québec, and conference delegates.[6]

JAMES BAY AND NORTHERN QUÉBEC AGREEMENT

The La Grande complex, also referred to as James Bay 1, was a fourteen-billion-dollar (Canadian dollars) project that produced ten thousand megawatts of electricity. It never had undergone an environmental assessment, for the simple reason that the impact assessment and review procedures did not exist when the project was announced in the early 1970s. The most significant impact of this particular project, without question, however, was the James Bay and Northern Québec Agreement (JBNQA), which was signed in 1975 by the government of Québec, the government of Canada, Hydro-Québec, the Crees, and the Inuit.[7]

As the first modern treaty signed in Canada, the JBNQA established environmental and social protection mechanisms involving full participation by the Native peoples. In a nutshell, the treaty exchanged all Cree and Inuit rights in the territory covered by the agreement (over one million square kilometers, or four hundred thousand square miles) for specific rights and granted the two groups monetary compensation in the amount of $225 million Canadian dollars (up to $500 million, when subsequent complementary agreements are factored in). The JBNQ recognized the Native peoples' right to harvest, set up a permanent financial program to support traditional harvesting activities, and provided for the participation of the Crees and the

Inuit in the environmental protection regimes. It also established the first Native self-government structures in Canada, including the Cree Regional Authority, the Cree School Board, and the Cree Regional Board of Health Services and Social Services; and, north of the 55th parallel, the Kativik Regional Government, the Kativik School Board, and the Kativik Health and Social Services Council, for the Inuit.

Although the Crees and the Inuit relinquished all their rights and Native claims in or to the lands of northern Québec, the JBNQA recognizes their exclusive trapping rights on the entire territory (1,066,000 square kilometers, or 410,000 square miles), the right to fish and hunt at all times, exclusive use of certain species (e.g., whitefish, black and polar bears), and exclusive harvesting rights on 162,324 square kilometers (62,678 square miles) of Category II lands. The Inuit Hunters Support Program and the Income Security Program for Cree Hunters and Trappers are intended to insure that traditional harvesting activities constitute a viable way of life for interested Native peoples, by providing them a measure of economic security.

These programs have been among the most successful aspects of the JBNQA. They have enabled the Native people to preserve their way of life (over 35 percent of the Crees still pursue traditional activities) and maintain a high proportion of traditional food in their diet (50 percent of their food still comes from the land), despite the drop in market prices for fur and the increased cost of pursuing traditional activities.

Empowering the Crees and the Inuit to take charge of their local, municipal, educational, and health services, combined with financial outlays by the Crees and Québec and Canadian governments, has changed the nature and landscape of the Native communities considerably (new housing, municipal services, community centers, new schools, and health clinics in every community). The Crees and the Inuit of northern Québec now have the tools they need to gain access to, and participate in, the mainstream economy of our country, while being able to preserve their hunting, fishing, and trapping way of life. Today the Cree and Inuit nations are political forces in both Québec and Canada, as well as being players on the international scene. For example, the James Bay Crees are the first and only Native organization from Canada to belong to the United Nations Working Group on Indigenous Peoples.

Enforcement of the JBNQA has not been easy. Its imprecise wording leaves room for interpretation, and the lack of clearly defined government action has led to several court and political battles. For example, the Crees continue to believe that Hydro-Québec must obtain their consent for the construction of Great Whale, also known as

James Bay 2, while the Québec utility thinks it merely requires authorization in relation to the environmental protection regimes in which the Crees have full participation. Despite normal difficulties inherent in implementing a 625-page legal document, the seventeen-year saga of the JBNQA still can be considered a success.

Since the signing of the JBNQA, all new projects must undergo an environmental assessment, and every hydroelectric project carried out since James Bay 1 has been approved by the Crees in the context of complementary agreements entered into with Hydro-Québec. Those agreements address issues such as financial compensation and the mitigation of social and environmental impacts.

PUBLIC REVIEW OF GREAT WHALE

Only later did Hydro-Québec become subject to the IRP process, considered by the Great Whale Forum to be one of the best methods available for producing environmental impact statements. Guidelines for the environmental impact statement for the Great Whale hydroelectric project were submitted to Hydro-Québec by the Evaluating Committee in September, 1992.[8] They required the utility to consider all demand- and supply-side resources as alternatives for meeting new demand. Demand-side resources include various strategies for load management and conservation. Supply-side resources include hydroelectricity, thermal energy, purchases from private producers (including cogeneration), and other technologies, such as wind, solar, and geothermal energy.

The guidelines suggested by the Great Whale Forum and other groups stipulated that "the Proponent shall carry out a least-cost analysis regarding the selection of supply- and demand-side resources to meet its load forecasts."[9] And they further stipulated that "the Proponent shall [also] present a series of tables comparing the impacts of each supply- and demand-side resource discussed in paragraphs 223–51 for each potential positive and negative externality. To the greatest extent possible, the Proponent shall make this comparison on a monetary or quantitative basis; however, for those impacts which do not lend themselves to quantification, the comparison shall remain qualitative."[10]

The environmental review of the Great Whale hydroelectric complex therefore was carried out using the IRP process. The review posed a considerable challenge for Hydro-Québec, as it forced the public utility to modify the planning and review process it normally applied to development projects.

The public review of the Great Whale project was led by the governments of Québec and Canada, in conjunction with the Native environmental committees. In January, 1992, the governments of Québec and Canada, Hydro-Québec, the Crees, and the Inuit signed a memorandum of understanding defining the environmental impact assessment procedures. However, the public review was brought to a halt in November, 1994, when Québec's Premier Jacques Parizeau announced that the project was being put on ice indefinitely (see below).

The public review of Great Whale set numerous precedents in Québec. Among them:

- This was the first time a hydroelectric project had undergone a comprehensive public review.
- The project assessment was based upon respect for the right of future generations (the local population and Québec citizens as a whole) to the sustainable use of the ecosystems in the Hudson Bay region.
- The notion of "public" was not limited to the Native population of northern Québec but included the entire population of Québec, Canada, and the U.S. In addition, the public could obtain financial assistance from the government to prepare briefs for, and participate in, the review.
- The guidelines for the environmental impact statement (EIS) were prepared by committees following public hearings held throughout Québec, with the participation in the hearings of Canadian and American organizations.
- Beginning in November, 1993, the committees also held public consultations to help determine the acceptability of the EIS submitted by Hydro-Québec in September, 1993 to comply with the guidelines issued in September, 1992.
- Final public hearings had to be held before the committees made their recommendations as to whether or not the development should go ahead and, if so, on what terms and conditions.
- The proponent had to demonstrate that there was a need for new or additional electrical generating capacity and that the best scenario for meeting that need included the proposed project. It had to do so using the methods of integrated resource planning and least-cost analysis. Alternatives to the project included the construction of new generating capacity, such as windmill farms and cogeneration stations, as well as programs to reduce demand growth.

WEAKNESSES OF THE GREAT WHALE PUBLIC REVIEW PROCESS

The Great Whale hydroelectric project underwent what was, without question, the most comprehensive public review ever undertaken of any North American development project. In hindsight, however, this review clearly had two major weaknesses. First, no deadline was set for completion of the environmental impact statements, which ended up taking inordinate amounts of time. Five years seems excessively long for the assessment of a development project.

Second, there appears to have been no logic to the order in which the public review was carried out. I am referring to the fact that, since there never had been a debate on energy in Québec, the section of the impact statements dealing with project justification turned into a broader debate on Québec's energy policy. Logically, the choice of hydroelectricity as a power source should have been justified well before modifications and mitigation measures for the Great Whale project were even contemplated. Instead, the opposite occurred, proving that logic is not always the prerogative of public decision makers!

SEPTEMBER, 1994: CHANGING COURSE

Following its electoral victory on September 12, 1994, the new Parti québécois government lost no time in making two crucial decisions with respect to Great Whale, or James Bay 2, as it was known.

First, in a surprising turn of events, Premier Jacques Parizeau announced in November, 1994, that Great Whale was being shelved indefinitely. There were several political reasons for this decision, but the two main ones were the projected lack of demand for electricity in the next ten years and the concern that the campaign against Great Whale would escalate and become a campaign against Québec in general, triggering any number of possible negative consequences. In the second critical decision, the minister of natural resources launched a vast public debate on energy, something that should have been undertaken years ago, while the Great Whale project was still on the drawing board.[11]

Thus the decision to shelve Great Whale indefinitely was dictated by two main imperatives:

- By 1995, Hydro-Québec, the public utility, now had a surplus generating capacity of four thousand megawatts.
- The need to assess Québec's energy policy as a whole—especially the role of the various forms of energy, including hydroelectric power—meant that any assessment of a project the size of Great Whale had to be put on hold. In the meantime, thanks in large part to the postponement of Great Whale, Québec and the James Bay Crees confirmed their will to reopen discussions, through an agreement signed by the Cree chiefs, the Québec premier, and myself in May, 1995.

The far-reaching energy debate held in 1995 made possible, for the first time in Québec, a collaborative process that brought together partners from a single sector—namely, energy—to define a policy that would guide the development of that sector. This process was inspired by the California experience in the early 1990s, in which a collaborative process was used to draft the state's energy policy. The consultation process is led by the actual players from Québec's energy industry, who subsequently will be responsible for defining the general aims of the government's new energy policy. Hydro-Québec, Greenpeace, Native Peoples, and other groups are working together to build a better future for all Québec citizens, instead of engaging in a futile debate through the media.

FUTURE PROSPECTS

Americans, Canadians, and Quebecers alike depend on a power transmission network that is part of a changing market characterized by numerous economic exchanges, spot markets, and aggressive competition. Also, we all are part of the same economic bloc governed by NAFTA, and thus our increasing interdependence as trading associates is inevitable. As economic partners, it is our responsibility to plan the foreseeable future together.

In my opinion, this future includes:

- Increased exchange of electricity between Québec and American utilities.
- Strong pressure on Québec's electricity market to open up and make greater room for independent power producers.
- U.S. independent power producers' opposition to Québec's presence in the U.S. market without a reciprocal guarantee—that is, the right to sell electrical power on the Québec market subject to the same rules.
- Opening up and decompartmentalization of Québec's electricity market.
- A political will on the part of those states tied to the same exchange network to encourage fair competition, primarily in terms of a transmission pricing formula established by using the integrated resource planning method to determine savings.

Given these conditions, it would not be at all surprising to see the creation of a regional transmission group composed of representatives from all bodies in the northeastern U.S. and, of course, Québec, in a bid to harmonize the electricity market within our joint territory. This would be one means of respecting the constitutional and legislative autonomy of each partner, while insuring that, whatever the means of generation, they would be compatible with the social and physical environment of our respective states. At the same time, we would be making sure that all partners consider and plan joint measures to mitigate transboundary impacts of power generation, in keeping with our willingness to act in partnership and maintain friendly relations.

NOTES

1. Hydro-Québec, *1994 Annual Report* (Montréal: 1995), 71.
2. New York Power Authority, press release, Mar. 27, 1992, in author's possession.
3. Editor's note: For a different point of view, see the *New York Times,* Jan. 12, 1992, 16; and the coverage on Mar. 17, B1: "The New York Assembly, in unusual foray into foreign policy, approves bill to require exhaustive environmental review before state purchases power from enormous hydroelectric complex that Hydro-Québec wants to build in remote James Bay." For an Inuit perspective, see Mary May Simon, *One Future— One Arctic* (Peterborough, Ont.: Cider Press, 1996), 45–48.
4. Editor's note: In 1993, Bouchard and other federal MPs from Québec bolted the Progressive Conservative Party of Brian Mulroney and formed the opposition Bloc Québécois in Ottawa. Following the Québec sovereignty referendum—which failed by less than 1 percent of the vote—in Oct., 1995, Bouchard returned to the province as premier and leader of the Parti Québécois, the provincial sovereignist party.
5. The firm was Écosystème PIR, Inc., and Cliche also was a partner in Groupe Consensus, Inc., a center for environmental and social mediation in Québec. Elected as a member of the National Assembly of Québec for the riding (district) of Vimont in the general election of Sept. 12, 1994, he then served as parliamentary assistant to the premier on native affairs before becoming Minister of Environment and Wildlife in the Bouchard government, and now Minister of Tourism.
6. These hearings were covered in the principal newspapers, including *Le Journal de Montréal, Le Devoir* (Montréal), *The Gazette* (Montréal), *La Presse* (Montréal), and *Le Soleil* (Québec City). See also Nicole Giasson, *Forum Grande-Baleine: L'examen du complexe Grande-Baleine,* papers of a conference held on June 11, 1991, in Québec City (unpublished document); *Forum québécois pour l'examen public du complexe Grande-Baleine,* 2nd conference, Library of Hydro-Québec. *Forum Grande-Baleine: Connaissance du milieu bio-physique et humain de la région de la Grande rivière de la Baleine,* papers of a conference held at Laval University, Québec City, Dec. 3 and 4, 1991 (unpublished document). Proceedings of the June 11 colloquium, edited by Nicole Giasson, are included in this second document.
7. Province of Québec, Secrétariat aux affaires autochtones of the ministère du Conseil exécutif, *James Bay and Northern Québec Agreement and Complementary Agreements* (Québec City: Les Publications du Québec, 1991 ed.).
8. Evaluating Committee, Kativik Environmental Quality Commission, Fed-

eral Review Committee, and Federal Environmental Assessment Review, *Guidelines: Environmental Impact Statement for the Proposed Great Whale River Hydroelectric Project* (Montréal: Great Whale Review Support Office, 1992). See also Hydro-Québec, *Complexe Grande-Baleine: Rapport d'avant projet—résumé* (Québec City: Hydro-Québec, 1983).

9. Evaluating Committee, *Guidelines,* para. 206.
10. Ibid., para. 211.
11. The public debate over energy, the prelude to a new energy policy, was launched by Minister of Natural Resources François Gendron on Feb. 7, 1995. Government of Québec, press release.

Part 2

The South: Mexico–United States Case Studies

5

MANAGING AIR QUALITY IN THE PASO DEL NORTE REGION

PETER M. EMERSON, CARLOS F. ANGULO,
CHRISTINE L. SHAVER, AND CARLOS A. RINCÓN

Along the U.S.-Mexico border, environmental conditions reflect the strains of poverty and heavy resource use. Fortunately, there also exist a persistent spirit of self-reliance and many examples of community leaders and citizens from both countries working together to achieve a better future. One of the most interesting community-led efforts for environmental improvement is the development of a cooperative strategy for transboundary management of air quality in the Paso del Norte region.

After serving as an important but remote crossroads of trade for centuries, the Paso del Norte region today hosts an international metropolitan complex of about 1.8 million people, most of whom live and work in the sister cities of Ciudad Juárez, Chihuahua; El Paso, Texas; and Sunland Park, New Mexico.[1] Population and industrial activity continue to grow. Occupying a wide valley surrounded by mountains, residents of Paso del Norte share a single air basin in which airborne pollutants are readily transported back and forth across the international border. For many years, the air basin has recorded unhealthy levels of ozone, carbon monoxide, and particulate matter in excess of national standards in both the United States (U.S.) and Mexico. Exposure to these pollutants can cause adverse health effects, such as: asthma attacks, bronchitis, chest pains, headaches, shortness of breath, irritation of the eyes and lungs, reduced attention span, diminished mental skills, and greater susceptibility to communicable diseases.

Local, state, and federal agencies in both countries have worked diligently to find ways to address the regional air pollution problem.

Under the 1983 U.S.-Mexico Border Environment Agreement, significant progress has been made in holding binational planning meetings and in conducting several air pollution studies.[2] For many years, however, incomplete scientific information, huge economic disparities, and national sovereignty considerations thwarted the design and implementation of a transboundary air quality management program.

In late 1992, the debate over the likely environmental consequences of the North American Free Trade Agreement (NAFTA) created a new opportunity to focus public attention on the long-standing need to reduce air pollution in the Paso del Norte region. In response to this opportunity, the Paso del Norte Air Quality Task Force—a binational group of business leaders, regulators, scientists, environmentalists, and elected officials—was organized to work for cleaner air.[3] Through its regularly scheduled meetings (the first of which was held on May 20, 1993) and specific pollution reduction projects, the task force has helped the citizens of the region reach agreement on the value of cleaner air and take responsibility for achieving it.

But local cooperation could not achieve its potential so long as national air quality officials continued to define air quality goals and programs strictly on a national basis, ending at the international border. On May 7, 1996, the U.S. and Mexican governments signed a pioneering agreement empowering the people of Paso del Norte to develop cooperative transboundary strategies to improve air quality throughout the air basin and committing the two governments to implement those strategies through national law. As a result, private citizens living in the region now have a legally recognized mechanism to engage government regulators and to work as a binational community on joint management of the shared air basin.

This paper has four major sections. The first section describes the air pollution problem in the Paso del Norte region. The second section sketches the relevant environmental statutes, international agreements, and administrative authorities of Mexico and the U.S. that relate to transboundary air quality management. The third section describes the steps that were taken to create an international air quality management basin. The fourth section offers recommendations for air pollution control in the Paso del Norte region in particular, and for transboundary resource management in general.

THE REGIONAL AIR POLLUTION PROBLEM

The sister cities of Ciudad Juárez, El Paso, and Sunland Park are located in a mountain pass formed where the Rio Grande, or Rio Bravo, intersects the Rocky Mountains in the Chihuahuan desert. The cities share a single air basin that has a base elevation approximately 3,700 feet above sea level. The air basin is defined by the Hueco Mountains in Texas to the east, the Juárez Mountains in Chihuahua to the south, the Franklin Mountains in Texas to the west, and tablelands that gradually rise to the Organ Mountains in New Mexico to the north.[4] Wind rose data indicate that Ciudad Juárez is downwind from El Paso and Sunland Park roughly 20 percent of the time, while the reverse is true roughly 15 percent of the time.[5]

In this complex mountain terrain, stable air masses and radiation inversions occur regularly during the fall and winter months. Under such conditions, cooler pollution-laden air is trapped near the ground beneath a layer of warm air and forms a "brown cloud" that can be seen in the valley. During other times of the year, hot, sunny days are quite common, creating ideal conditions for photochemical reactions that turn certain air contaminants into ozone pollution. Semiarid conditions of no more than seven or eight inches of rainfall per year and frequent winds are responsible for the entrainment of much particulate matter.

Air quality monitoring programs have been in place in the Paso del Norte region since the 1970s. El Paso air quality exceeds several U.S. national air quality standards; different parts of the city have been designated as a serious nonattainment area for ozone, and moderate nonattainment areas for carbon monoxide and particulate matter.[6] Air quality monitoring in Ciudad Juárez indicates a pollution problem at least as severe as that in El Paso. Ambient concentrations in Ciudad Juárez regularly exceed Mexican air quality goals for ozone, carbon monoxide, and particulate matter.[7]

Ozone is a major component of the haze, or smog, that troubles many large cities. It is not discharged directly into the atmosphere by polluters, but is formed in the air. Volatile organic compounds and nitrogen oxides that are emitted from many sources—in the Paso del Norte air basin, primarily from vehicles—undergo photochemical reactions in the presence of sunlight, causing oxygen to become ozone. In El Paso, the federal ozone standard is typically exceeded five to seven days per year in the summer or early fall.[8] Statistical analysis shows that the trend in peak ozone concentration level was upward before 1989 but has decreased slightly since that time.[9]

Monitoring data for El Paso and Ciudad Juárez reveal elevated levels of carbon monoxide in the fall and winter months, with the highest levels often occurring in December on days of low wind speed. From 1981 to 1990, the federal carbon monoxide standard was exceeded, on average, twelve days each year.[10] El Paso currently is the only area in Texas to exceed this standard. Data for the winter months of 1992–95 show a marked downward trend in the magnitude and frequency of carbon monoxide exceedences in El Paso. However, for the years 1982–93, data on the highest readings for carbon monoxide from an El Paso monitoring site do not indicate a downward trend.[11] Carbon monoxide measurements—which also have been high in June and July—are known to vary widely, depending on atmospheric conditions.

High levels of inhalable particulate matter (called PM-10, to denote a particle size of ten microns or less) generally are recorded on stagnant weather days during the fall and winter months, when particles are trapped near the ground. They tend to occur in the downtown business areas of El Paso and Ciudad Juárez and often are associated with visible air pollution and the region's brown cloud. El Paso is the only area in Texas that exceeds both the federal twenty-four-hour standard and the annual standard for particulate matter.[12] Exceedences above the twenty-four-hour standard regularly are observed at monitoring sites on both sides of the border, although the highest readings of particulate matter in El Paso have shown a decreasing trend in recent years.[13] Several years ago, it was determined that levels of benzene, a known carcinogen that is found in particulate matter, were 3.5 parts per million (ppm) in El Paso and 28.8 ppm in Juárez.[14] The American Lung Association has warned residents of both cities that they face serious health threats from respiratory disease as a result of particulate matter pollution.

Air pollution in the Paso del Norte region comes from many sources, manmade and natural, and from both sides of the border. Its accumulation, movement back and forth across the border, and adverse impacts are intensified by the unique topography and meteorology of the natural air basin. Using information from the emissions inventory for El Paso and Sunland Park and from studies that have been carried out in Ciudad Juárez, air quality experts believe that major sources of pollutants in the Paso del Norte air basin include: motor vehicles; open-burning of trash; home fuel consumption; fuel transport and storage; dust from highway traffic, construction materials, and equipment; brick ovens and small-scale industrial sources; and fugitive solvents from painting, architectural coatings, and manufacturing processes.[15] Heavy industry in the region—a copper smelter, two refineries, and a large cement plant—also contributes pollutants.

Over the past decade, the number of sources emitting pollutants in the Paso del Norte air basin has increased rapidly, as the populations of Ciudad Juárez and El Paso grew by 40 percent and 23 percent, respectively.[16] However, the important question of which political jurisdiction contributes the most air pollution in the air basin cannot be answered precisely. It is known that Ciudad Juárez has fewer pollution controls, a much larger population, a significantly older vehicle fleet, many more miles of unpaved roads, and more open-burning than El Paso and Sunland Park. Certainly, each of these factors contributes large amounts of air pollution to the total originating in Ciudad Juárez. On the other hand, resource utilization tends to be greater in higher-income El Paso and Sunland Park; this in turn is responsible for large amounts of air pollution. Some environmental benefits of newer, cleaner-running vehicles and paved streets in the U.S. cities are partially or completely offset by the fact that El Paso and Sunland Park residents have more cars, and their vehicle mileage per capita is about six times higher than in Ciudad Juárez.[17]

Industrial growth in the Paso del Norte region has been driven by a boom in the number of *maquiladora* (assembly) plants located in Ciudad Juárez and northern Mexico. Serving customers in the U.S. and other markets, production processes at these plants—especially those manufacturing automotive parts and electrical parts, which use large amounts of solvents—contribute to air pollution. The establishment of the plants also has triggered a huge increase in industrial traffic. For example, "in 1988 about 500 trucks crossed the international boundary (in the Paso del Norte region) each day; in 1993 it was about 2,000 trucks daily."[18] In 1995, there were 1.2 million northbound international bridge crossings by commercial vehicles in the region.[19] Heavy transport vehicles and traffic stalled at the border crossings are an important source of air pollution emissions.

Scientists agree that the air pollution problem in the Paso del Norte region is transboundary with respect to both its origin and its adverse impact on human health and the environment. To understand fully the magnitude and causes of the problem, U.S. and Mexican scientists and government officials need to complete the creation of an air quality monitoring network and an emissions inventory for the air basin. They also need to continue working together on pollution characterization studies and modeling efforts. Development of an effective transboundary management strategy to reduce air pollution will require both countries to adopt policies and implement mitigation tools that meet the special needs of the international air basin.

CURRENT EFFORTS TO REGULATE AIR QUALITY

Management of the environment along the U.S.-Mexico border is shaped by the domestic policies and priorities of the two countries and by their international commitments. This section describes existing environmental statutes, international agreements, and administrative authorities of Mexico and the U.S. that regulate transboundary air quality in the Paso del Norte region.

Mexican Air Quality Regulation

In Mexico, the development and implementation of environmental policy generally is highly centralized, with virtually all powers falling under the federal government's jurisdiction. However, in recent years, efforts have been made to decentralize environmental policy responsibility to states and municipalities.

Article 27 of the Mexican Constitution provides the main legal framework related to natural resources. Its first paragraph deals with matters related to private property regulations and restrictions to protect the public interest and provides that the federal government will adopt the measures necessary to regulate land use and to preserve and protect the ecological balance. *Ecological balance* is defined by the 1988 General Law for Ecological Equilibrium and Protection of the Environment (referred to here as "the Environmental Law") as "the relationship of interdependence between the elements that compromise the environment, which makes possible the existence, transformation, and development of mankind and other living beings."[20]

Article 73, Section 29-G, of the Mexican Constitution states that the Mexican Congress has the power to issue laws related to environmental protection and preservation of the ecological balance. Congress has exercised this authority in recent years, enacting a number of important laws, regulations, and standards, the most important being the 1988 Environmental Law. Currently, the administration of President Ernesto Zedillo Ponce de Leon is continuing a thorough review of the Environmental Law, with the intent of proposing a comprehensive set of amendments to that law. Although the administration has released several proposed amendments, it is difficult to predict what changes ultimately may be adopted. While the provisions on air quality addressing allocation of jurisdiction may be modified, no significant change in the general objectives of the law is expected.

The Environmental Law was the result of the first serious effort on the part of the Mexican government to regulate a number of environmental topics, such as natural protected lands, environmental impact, hazardous activities and hazardous substances, and prevention and control of soil, water, and air pollution. The Environmental Law also establishes administrative and criminal sanctions, and outlines which areas of environmental legislation fall within the jurisdiction of federal, state, or municipal regulating agencies.

The Environmental Law provides that the federal government has jurisdiction over natural protected areas, activities considered to be highly hazardous, all matters related to hazardous materials and hazardous waste, prevention and control of water pollution when wastewater is discharged into federal bodies of water, and the prevention and control of air pollution when air emissions affect two or more states or another country. The Environment Law also has a blanket provision stating that the federal authority has jurisdiction over "those matters which due to their nature and complexity require the federal government's participation."

Since 1988, several regulations have been promulgated dealing with environmental matters, most of which derive from the Environmental Law. On November 25, 1988, regulations in the area of prevention and control of air pollution were published. These regulations state that authority in air quality issues shall be exercised concurrently by federal, state, and municipal authorities and basically repeat the provisions of the Environmental Law related to air emissions affecting two or more states or another country.

Finally, the federal government has issued a great number of environmental standards, known as "Official Mexican Standards" (Normas Oficiales Mexicanas, or "NOMs"), including NOMs on permissible levels of air pollution. These NOMs play an important role in the actual application and enforcement of federal laws and regulations, because their purpose is to establish technical and scientific guidelines to be applied in the protection of the environment. In the area of air quality, among the most important NOMs are those setting maximum air emission levels for solid particles from fixed sources; maximum air emission levels of pollutant gases from vehicles that use gasoline as fuel; maximum allowable levels for particles, carbon monoxide, nitrogen oxide, sulphur dioxide, sulfuric acid fog, and suspended particles; and conditions and requirements for the operation of equipment that uses direct heating through combustion (that is, fixed sources utilizing liquid or gas fossil fuels).

The majority of Mexico's thirty-one states also have enacted their own environmental laws since 1988. Many of these laws are quite similar to those of their federal counterpart and contain provisions dealing with state environmental policy, including air pollution. In the case of state environmental laws and regulations, federal NOMs are also the applicable technical standards used by state authorities in enforcing such laws and regulations. The State of Chihuahua published its own environmental law ("Ecological Law of the State of Chihuahua") on October 26, 1991. This law contains a specific chapter dealing with prevention and control of air emissions originating from private motor vehicles and fixed sources under state or municipal jurisdiction.

Mexico's federal Environmental Law and regulations provide two means for enforcing air pollution control requirements: sanctions and operating licenses. Sanctions include: plant closures (temporary or permanent; partial or full); administrative arrest; imposition of fines; and revocation or cancellation of a source's license. Criminal sanctions also are authorized for flagrant and criminal violations.

The environmental regulations include a licensing requirement and process for stationary sources. Licenses generally set maximum permissible emission levels consistent with the NOMs, but authority exists to establish source-specific limits. In addition, allowable emissions specified in an operating license may be modified if the source is located in a "critical zone," such as the Northern Border Area (defined as the strip one hundred kilometers wide along the border). Licenses also may be modified if a more efficient technique exists to control air pollution or if production processes change at the source. Therefore, Mexican law and regulations provide ample authority for establishing, modifying, and enforcing emission limits on stationary sources.

U.S. Air Quality Regulation

In the U.S., air pollution is regulated under the federal Clean Air Act, which most recently was amended in 1997. The Clean Air Act is one of the most complex environmental laws in the U.S., and there are numerous federal regulations and guidelines that go into even greater detail on many aspects of its implementation.[21] The following summary is intended to give only broad outlines; therefore, it necessarily oversimplifies details that may be important for individual sources or programs.

Since 1970, the U.S. has followed an air pollution control strategy that focuses on the universal attainment of National Ambient Air Quality Standards (NAAQS) for six common air pollutants: particulate matter, sulphur dioxide, carbon monoxide, nitrogen oxide, ozone, and lead. The NAAQS are set at the federal level, according to scientific criteria, in order to protect the public health with a margin of safety. Typically they include standards for both short-term peak concentrations and long-term average concentrations. Geographical regions that have not yet attained the NAAQS for a pollutant are termed *nonattainment areas.*

As noted earlier, El Paso is a nonattainment area for ozone, carbon monoxide, and particulate matter. Although many U.S. cities exceed the ozone standard, El Paso is one of only a few cities that is nonattainment for three separate pollutants. All statutory deadlines for attainment of standards expired at the end of 1996, except for cities with serious or severe ozone, carbon monoxide, or particulate matter problems.

Under U.S. law, a nonattainment area is subject to stringent cleanup requirements and may be penalized for failure to meet the requirements. A border city like El Paso, however, cannot be penalized if it shows that it has taken all necessary measures within its jurisdiction and that its failure to meet the requirements is due only to emissions outside the U.S. (section 179B of the Clean Air Act, providing this exemption, was added in 1990).

To attain acceptable air quality requires, of course, reduction of pollutant emissions into the atmosphere. The federal act divides sources of air pollution into two broad groups: mobile sources and stationary sources. For mobile sources, the emissions standards for motor vehicles and the fuel quality requirements for motor gasoline (e.g., no lead) are set at the federal level. In addition, the federal law calls upon the states and localities that are nonattainment for ozone to adopt various "transportation control plan" measures to reduce the use of cars and trucks, with the exact measures required depending in part on the severity of the local ozone problem. Finally, ozone nonattainment areas may be required to establish federally approved programs for vehicle emissions inspection and maintenance programs, but the exact design of such programs is left to the states and localities. Because El Paso's ozone problem is classified as "serious," its extended deadline for attaining that standard is November 15, 1999. Toward that goal, the area was required to achieve a 15 percent reduction in volatile organic compound emissions by the end of 1996 and a further 3 percent per year reduction, using the various strategies specified in the statute.

Under the Clean Air Act, the stationary source group is subdivided into two classes. For new sources or substantially renovated sources in certain major industrial classes, the U.S. Environmental Protection Agency (EPA) has established nationwide "new source performance standards" (NSPS) that set minimum pollution control requirements based on performance of the best control technologies. State and local circumstances, however, may require additional pollution controls beyond the NSPS requirements for some sources. Requirements for control of certain listed hazardous air pollutants also are set at the national level for existing as well as new sources, under a similar technology-based approach and evaluation of any residual health risks.

All existing stationary sources of air pollution and new sources in industry classes for which there are no NSPS are governed by a more complex array of controls implemented at the state level, though still under federal guidelines and minimum requirements. In general, it is up to each state to develop a "state implementation plan" (SIP) that will result in attainment of all air quality standards. The federal requirements for SIPs are set out in Section 110 of the act. The SIP will set minimum emissions control requirements for sources in many different industry classes and may set different requirements for different geographical areas, depending on current air quality and the degree of improvement that is needed. The state laws and rules must meet minimum federal requirements, but the states are free to adopt laws or rules that are more stringent or more comprehensive. Thus, for example, the State of Texas regulates many small sources not covered by the federal act, and regulates pollution problems (such as odors) that also are not covered under federal law.

The state clean air laws and regulations of Texas and New Mexico generally follow the federal scheme and provide detailed rules designed to implement the federal requirements at the state level. The Texas Clean Air Act contains the general legal authority for matters such as: design of transportation control plans, and inspection and maintenance programs for vehicle emissions control; the review and permitting of new sources; and the application of reasonably available control technology for existing sources. The Texas Natural Resource Conservation Commission (TNRCC) is given the authority for implementing the act's provisions through promulgation and administration of detailed administrative regulations that appear in the Texas Administrative Code. Notably, some aspects of the Texas statute and rules go beyond the federal act's minimum requirements. These provisions, such as control of outdoor burning or requirements for paving of parking lots and roadways in El Paso, represent the state's choices about how best to meet the national ambient air quality standards. It is also notable that the Texas Clean Air Act specifically instructs the TNRCC to implement fully all the federal requirements for stationary source control in El Paso, so that the state qualifies for the special exemption from sanctions for not meeting clean air deadlines, provided in section 179B of the U.S. Clean Air Act.

With respect to the process of emissions offsets and emissions banking, the Texas law and regulations provide for emission reduction credits, not only for stationary sources but for mobile sources as well. Thus, accelerated implementation of programs, such as switching El Paso's bus fleet to alternative fuels, can result in emission reduction credits usable in transactions with other contributors to the ozone nonattainment problem. The Texas rules also specify the use of oxygenated fuels in El Paso in the winter season to help reduce carbon monoxide emissions, and ban the use of wood-burning stoves and similar devices during inversion episodes to reduce particulate emissions.

In New Mexico, the controlling state statute is the Air Quality Control Act, administered by the state's Department of the Environment. The state's regulations appear in the New Mexico Administrative Code as a compendium of Air Quality Control Regulations. As with Texas, the basic provisions implement the federal act's requirements with respect to such matters as permits, review of new sources, transportation control planning in ozone nonattainment areas, and emission offsets.

For many years under Texas law, and federally required since 1990, each air-pollution source must have a permit that specifies the pollution control requirements for that particular source. The permit is the basic mechanism for identifying sources, setting limits, monitoring performance, and enforcing requirements for that source.

In nonattainment areas, the federal act specifies that the states shall insure that existing sources apply at least "reasonably available control technology" (RACT), as determined for each source individually by applying EPA technology guidelines. For inhalable particulates, RACT guidelines are supplemented by "reasonably available control measures" for area sources of particulates, such as roads and wastewater treatment ponds. For nonattainment areas, the federal Clean Air Act also requires that SIP provisions result in "reasonable further progress" toward attainment and that they include a procedure for reviewing the air quality effects and pollution control systems on any new facility or expansion of an existing facility emitting more than minimal amounts of the nonattainment pollutants in the area. (For ozone nonattainment areas, stationary sources of volatile organic compounds are regulated.)

Three federal Clean Air Act requirements must be met before a new or expanded facility can get a permit to operate in a nonattainment area. First, the permit applicant must show that other already existing sources of the same pollutant in the area will "offset" the new emissions with even greater reductions in current emissions (i.e., more than one ton per year of reductions for each ton per year of new permitted emissions). Second, the permit applicant must adopt the most stringent pollution controls at the new or expanded facility—referred to by the term "lowest achievable emissions rate." Third, the applicant must certify that all other facilities under its control in the state are complying with their air pollution control requirements.

To promote progress toward attainment by reducing emissions from existing sources and to facilitate the emissions trades that may allow new or expanded industrial activity in nonattainment areas, EPA rules allow sources to "bank" their credits for emissions reductions. The source can then draw on its "bank account" for offsets to its own increases in emissions, or it can sell those credits to a third party that may need offsets. Government agencies which generate air emissions or can show reductions of emissions through government programs also can contribute their emission reductions to private projects. For example, control of emissions from garbage incineration can reduce particulate matter, and changes in the specifications for road paving materials may reduce emissions of volatile organic compounds.

In the Paso del Norte region, the availability of emission offsets as both an opportunity for new facilities to locate in the region and as an incentive for further reductions in emissions from existing facilities has been constrained by the fact that Clean Air Act jurisdiction, and thus the opportunity to trade emissions, ends at the Rio Bravo/Rio Grande even though the air passes freely back and forth across the border. An important objective in creating the Paso del Norte Air Quality Management Basin is to make the benefits of offsets and emissions trading available to sources throughout the common airshed.

Binational Cooperation on Air Quality

For many decades, both U.S. and Mexican governments have recognized that environmental problems in the border region cannot be addressed effectively by different and uncoordinated approaches on each side of the border. In 1983, after trying several largely unsatisfactory ad hoc arrangements dealing with specific problems, the two countries entered into an "Agreement Between the United States of America and the United Mexican States on Cooperation for the Protection and Improvement of the Environment in the Border Area," known simply as the La Paz Agreement, after the city of La Paz, Baja California, in which it was signed. Air pollution was noted as one of the transborder problems that could be addressed under the agreement, which gives the two federal environmental agencies primary responsibility for environmental policy within one hundred kilometers on either side of the border.

In 1989, recognizing the serious nature of the air pollution problem in the Paso del Norte region and the need for binational cooperation to address the problem, the countries added Annex V to the La Paz Agreement. Annex V seeks a better understanding of the problem through a binational inventory of emissions sources, air quality monitoring, and air quality modeling. The studies conducted under Annex V have created a solid foundation for the next step—a coordinated, cooperative transboundary management strategy to improve air quality in the region by reducing emissions of air pollutants. Furthermore, when the U.S. Congress amended the federal Clean Air Act in 1990, it included specific authority (in section 815 of the amendments bill) for the executive branch to negotiate with Mexico on programs to monitor and improve air quality in the region.

Additional binational cooperation on air quality also has occurred through the U.S.-Mexico Integrated Border Environmental Plan, a broad program of cooperation between the Mexican and U.S. environmental agencies.[22] The border plan expanded the Annex V program of monitoring and modeling to include more emphasis on reducing vehicle emissions and pollution control strategy development. Recently, the governments have developed the elements of a second-generation plan called the U.S.-Mexico Border XXI Program.[23]

Some further international support for air quality management in the Paso del Norte region may come from the assessments, policy analyses, and factual reports of the North American Commission for Environmental Cooperation. The commission was established in 1994 by the NAFTA partners to address environmental issues of common concern, including matters of transboundary pollution. Finally, another NAFTA-related institution, the Border Environment Cooperation Commission (BECC), may have a role to play in helping to develop and arrange financial support for specific environmental infrastructure projects, such as new solid waste disposal facilities, that could help reduce air pollution. BECC-certified projects become eligible for financing through the North American Development Bank.

Summary

Current efforts to reduce air pollution in the Paso del Norte region depend on air quality policies, standards, and enforcement procedures promulgated in Mexico and the U.S. and on binational efforts authorized under the 1983 La Paz Agreement and other bilateral or continentwide arrangements. Although the region's air pollution is shared by everyone living in the air basin, there are important differences between the Mexican and U.S. pollution control regimes. Statutes, regulations, and enforcement procedures in each country differ significantly because of different constitutional structures, legal traditions, and stages of economic development. Furthermore, the legal jurisdiction of each country ceases at the international border. Together, these factors lead to a "governance failure"—a gap between each government's regulations controlling air pollution emissions and the transboundary character of the polluted air.

Because of the governance failure, air pollution control in the shared Paso del Norte air basin is costly and ineffective. However, the community's interest in better air quality led local leaders to advocate a transboundary management strategy that emphasizes binational cooperation, innovation, and local responsibility in achieving cleaner air. The positive working relationships that exist in the Paso del Norte region can promote the evolution of existing U.S. and Mexican regulations and future economic growth to benefit the entire air basin.

TAKING THE NEXT STEP: TRANSBOUNDARY AIR QUALITY MANAGEMENT

Twenty-two years ago, John C. Ross, then El Paso city attorney, speaking at the First Binational Symposium on Air Pollution along the U.S.-Mexico Border, observed, "Air pollution by its very nature transcends the international boundary without regard to jurisdiction areas, and ... we must seek legal solutions which likewise can bridge the jurisdictional lines between our two countries."[24] At the same symposium, another U.S. attorney proposed creation of an International Environmental Air Pollution Commission,[25] and a highly regarded Mexican attorney proposed a bilateral treaty to deal with Paso del Norte air pollution, noting, however, that "it would be necessary to protect the sovereignty of each State to insure that it had the right to enforce standards within its own boundaries."[26]

Establishing the Framework

In 1989, three professors from the University of Texas at El Paso (UTEP) made the case for a locally controlled, binational program to manage the international air basin. They were the first to point out that economic instruments, such as emissions trading, could be used in the air basin to lower the cost of air pollution cleanup below that of a command-and-control system.[27] Recognizing that it was not realistic for the two countries immediately to establish uniform ambient standards for the air basin, the professors recommended that the countries take three steps: (1) implement an air quality accounting system; (2) allocate initial control responsibility based on existing standards through operating permits; and (3) develop emissions trading procedures for uniformly mixed assimilative pollutants, such as volatile organic compounds. They reasoned that such a program would insure progress by each country toward its own air quality goals, although pollution control costs would fall disproportionately on the country with the most stringent standards. To a large extent, the 1989

UTEP proposal has served as the motivating framework for the most recent efforts to achieve transboundary management of air quality in the Paso del Norte region.

The likelihood of achieving this goal has been greatly increased by the important work of the Binational Air Working Group. Created under the auspices of the U.S.-Mexico Border Environmental Agreement, the Binational Air Working Group has met annually since 1984. It is currently co-chaired by Dr. Adrian Fernández, Instituto Nacional de Ecología, Secretaría de Medio Ambiente, Recursos Naturales y Pesca, and David Howekamp, director, Air and Toxics Division, Region 9, EPA. This group of U.S. and Mexican scientists and air pollution control administrators, along with officials from Ciudad Juárez and El Paso, have planned and carried out projects on air quality monitoring, emissions inventories, particulate matter transport, vehicle emissions, air toxics, and modeling of pollution control strategies.

Obtaining Authority for Transboundary Management

In 1992, the NAFTA debate focused attention on environmental issues along the U.S.-Mexico border, creating a new opportunity to make headway on the Paso del Norte air pollution problem. In meetings with the late Luis Donaldo Colosio, Mexico's top environmental officer at the time, Environmental Defense Fund (EDF) staff members discussed several ways that new transboundary institutions and economic instruments could be used to solve problems along the border.[28] Colosio was interested in ideas that would help decentralize and increase the efficiency of environmental regulation in Mexico. He was especially supportive of the EDF's recommendation that an international air quality management district be created in the Paso del Norte region.[29] As a result, Colosio directed Dr. Miguel Angel Orozco Deza, who then served as delegate from the Procuraduría Federal de Protección Ambiental in Chihuahua, to work with the EDF staff in carrying out this recommendation.

With the encouragement of Mexican and U.S. officials, the EDF staff next pursued a series of meetings in Ciudad Juárez and El Paso. Important activities included a presentation to the Mayor's Environmental Roundtable in El Paso, work with air quality experts at the Dirección Municipal de Ecología and the City-County Health Environmental District, and meetings with local business leaders. In April, 1993, with the support of El Paso–Ciudad Juárez environmental officials and business leaders, the EDF staff appeared before the Texas Air Control Board (now reorganized as part of the Texas Natural Resource Conservation Commission) to recommend creation of a regional task force to help solve the Paso del Norte air pollution problem.

The board accepted the recommendation and appointed Texas members to the task force.

The first meeting of the Paso del Norte Air Quality Task Force was held on May 20, 1993. Chaired by Dr. Elaine Barron, an El Paso physician and a member of the Texas Air Control Board, the meeting was attended by about fifteen people. Today, the task force meets five or six times a year, with forty or more business leaders, regulators, scientists, environmentalists, and elected officials from both countries in attendance. The task force has worked very closely with key government officials, and both U.S. and Mexican members have asked for more local responsibility in air quality regulation. From the beginning, the effectiveness of the task force was enhanced by the active participation of Francisco Núñez, Ciudad Juárez's chief environmental officer, and Kirk P. Watson, chair of the Texas Air Control Board. After creating the task force, Watson took the lead in publicly supporting an innovative approach to transboundary air quality management.[30]

Throughout 1993 and 1994, task force members worked to understand the regional air pollution problem and to gain the confidence of the EPA, Mexico's federal environmental agency, and other government agencies. The task force also debated and prepared—with the assistance of attorneys from EDF, Ciudad Juárez, El Paso, TNRCC, Chihuahua, and the Texas General Land Office—a proposed Annex VI to the La Paz Agreement, creating an international air quality management district (IAQMD) for the Paso del Norte region (see appendix A).

The task force envisioned that the IAQMD would not supersede regulatory authority of the two countries, but that it would act at the local level to conduct joint activities in data gathering, pollution prevention, technology transfer, compliance review, public education, and development of transborder pollution control strategies, including the use of economic instruments such as an emissions trading program. The task force also recommended a governing board drawn from the responsible government agencies and interested citizens living in the region. After preparing its proposed Annex VI, the task force succeeded in winning recognition and support from many elected officials at all levels of government, including Presidents Clinton and Salinas and, later, President Zedillo.

As the proposed Annex VI was being reviewed by federal authorities, the task force continued to meet on a regular basis and was active in community outreach concerning the region's air quality and in carrying out specific pollution reduction projects. For example, EDF and the task force have worked with officials in Ciudad Juárez to improve the vehicle inspection and maintenance program by helping to set up vehicle emission diagnostic centers and a training program for automobile mechanics. The task force also has worked with government agencies on initiatives related to extending the U.S. Department of Energy's Clean Cities Program to Ciudad Juárez, speeding the introduction of alternative-fuel vehicles, and reducing traffic congestion on international bridge crossings. A project also was started to quantify and address paint and solvent shop emissions in Ciudad Juárez, including preparation of a handbook on reducing emissions.

The work on these pollution reduction projects has benefited from a successful project aimed at helping the owners of high-polluting brick kilns in Ciudad Juárez to adopt cleaner fuels and more energy-efficient kilns. Administered by the Mexican Federation of Private Associations for Health and Community Development, the Ciudad Juárez Brickmakers' Project (see next chapter in this book) has involved scientists from El Paso Natural Gas, and Los Alamos and Sandia National Laboratories. The Brickmakers' Project is a model of effective binational cooperation.[31]

In June, 1995, at a La Paz Agreement meeting in Mexico City, the two governments agreed to enter into formal negotiations regarding the establishment of an IAQMD and a joint committee on air quality improvement for the Paso del Norte air basin. In August, 1995, U.S. and Mexican officials met in Washington, D.C., for their first negotiating session. During the remainder of 1995, the federal negotiators continued to exchange information among themselves and with their respective state governments. Also a delegation of task force members from Ciudad Juárez and the EDF staff met with environmental and foreign relations officials in Mexico City, to be briefed on the negotiation process and to make specific recommendations to the negotiators.

On May 7, 1996, Secretary of State Warren Christopher and Secretary of Foreign Relations Angel Gurria signed an agreement adding a new Appendix 1 to Annex V. The appendix created a binational committee on air quality improvement for an international air basin defined as those parts of El Paso County, Texas; Doña Ana County, New Mexico; and the metropolitan area of Ciudad Juárez, Chihuahua, that are within one hundred kilometers of the border (see appendix B). The official title of the committee is Joint Advisory Committee on Air Quality Improvement for the El Paso–Ciudad Juárez–Doña Ana County Air Quality Management Basin (here referred to as the "Joint Committee"). Christopher praised the agreement as a unique cooperative approach to "enable business and government leaders from Texas, New Mexico, and Ciudad Juárez to reduce some of the region's worst air pollution."[32]

The Joint Committee will develop and present recommendations to the Binational Air Working Group on the prevention and control of air pollution in the international air quality management basin (IAQMB). According to a "guidance document" prepared by the negotiators, the Joint Committee's work is expected to include recommendations on:

- The integration of air quality monitoring networks and reporting of air quality pollution indices for the IAQMB.
- Exchange of information, technical training, and technology transfer to benefit air quality.
- Development and implementation of public education and outreach programs for the IAQMB.
- Use of air quality modeling to evaluate specific pollution abatement and prevention strategies.
- Development and implementation of economic instruments, including emissions trading programs (with emissions budgets, loading allowances, and/or caps), for the IAQMB.
- Specific steps to improve the effectiveness of air quality programs in the IAQMB.

Membership of the Joint Committee will consist of twenty persons, ten selected by each country. The representatives from each country will include five government officials from federal, state, and municipal jurisdictions and five nongovernment members from businesses, civic, and environmental organizations, and academia. The nongovernment members of the Joint Committee must reside in the jurisdiction of the IAQMB. Members will be appointed by the EPA and the Secretaría de Medio Ambiente, Recursos Naturales y Pesca; and all meetings must be open to the public. Otherwise, the Joint Committee will establish its own rules of procedures and meeting schedule.

ORGANIZING THE JOINT ADVISORY COMMITTEE ON AIR QUALITY IMPROVEMENT

With the international agreement now in place, the next step toward transboundary management of air quality in the Paso del Norte region is organizing the new Joint Advisory Committee. Implementation of the agreement needs to be carried out in a way that fully involves the local community and contributes to efficient and open operating procedures. It is important, too, that the Joint Committee achieve early improvements in air quality and its management throughout the basin.

Involving the Local Community

Both the U.S. and Mexico have agreed that nongovernment representation is very important to insure that the Joint Committee reflects and responds to local concerns, needs, and priorities. It is true that recommendations made by the Joint Committee will be forwarded to the Binational Air Working Group for a final decision, and many proposed actions will require federal or state government involvement. Still, those who live and work in the Paso del Norte community must have a significant role in developing cost-effective solutions to the air quality problems that will be implemented cooperatively at a local level. Therefore, commitment from the business communities and local governments will determine whether strategies are implemented successfully.

The Joint Committee should become a primary forum for receiving public input on air quality problems and solutions. It also must be accountable to the public and assume responsibility for disseminating information and reporting on actions taken and results achieved. In order to carry out these functions effectively, the Joint Committee will need to establish and maintain a good working relationship with the Binational Air Working Group, so that there is a routine method for acquiring necessary information and holding the federal and state governments accountable for actions taken in response to the Joint Committee's recommendations. The Joint Committee also will need to interact regularly with the Paso del Norte Air Quality Task Force. The task force members not only are knowledgeable about local air quality problems, but also are action-oriented and could play a significant role in tapping financial resources and skills needed to implement strategies recommended by the Joint Committee.

Operating Efficiently and Openly

In order to maintain credibility, the activities of the Joint Committee should be carried out in an open and transparent manner. Therefore, the Joint Committee's administrative procedures need to specify the frequency, time, and place (preferably alternating among the affected communities) of meetings; indicate that all meetings will be open to the public, with adequate notice provided and time set aside for public input; and explain where and how records may be accessed by the public.

Successful implementation of the Joint Committee's recommendations will depend upon commitments from various governmental officials, as well as the public and business community. Therefore, decision making within the Joint Committee should be done by consensus, with equal weight given to the views of all representatives.

Because formal meetings of the Joint Committee may occur only a few times each year, a method for routinely receiving public input and disseminating information needs to be established. The Joint Committee should have a continuous and known presence in the region. Although federal government officials will chair the Joint Committee, a local liaison office or person should be established. The role of the liaison could include maintaining records, responding to inquiries, and monitoring relevant activities of state and local governments as well as the Paso del Norte Air Quality Task Force. Similarly, because the Binational Air Working Group meets only a few times a year, it should establish a point of contact for the local Joint Committee liaison.

The operating procedures of the Committee should explain how the Joint Committee will interact with the Binational Air Working Group. The procedures should outline the Joint Committee's expectations regarding timely decisions on recommendations, as well as a process for appealing negative decisions, including the option of pursuing alternative means for implementation if no federal action is needed. Joint Committee meetings should be held in conjunction with Binational Air Working Group meetings, to facilitate timely response to recommendations. The operating procedures could outline methods for streamlining decision making and strategy implementation, particularly if the Joint Committee meets more frequently than the Binational Air Working Group. The Joint Committee could develop criteria for categorizing recommendations and propose alternative decision-making paths, so that routine recommendations could be implemented more expeditiously. For example, if a particular recommendation can be implemented by local governments or private interests and no additional federal funding is needed, the Joint Committee should be authorized to proceed without formal action by the Binational Air Working Group.

The Joint Committee should have responsibility for estimating the financial, human, and technological resources needed to implement its recommendations, as well as identifying the means to obtain necessary resources. This will help expedite decision making and stimulate a positive response from both governments.

ACHIEVING AIR QUALITY OBJECTIVES

The credibility of the Joint Committee will depend upon community participation and a fair and open process for conducting business, but its effectiveness will be judged by what it achieves. Therefore, an important first step for the Joint Committee will be establishing objectives and deciding how to achieve them. One of the highest priorities should be to agree on short- and long-term air quality objectives for the international air basin, consistent with attaining the respective national ambient air quality standards. Targets and timelines for air quality and emissions should be established as measures of progress.

Information Acquisition

Establishing and achieving air quality objectives will require reliable information. The Joint Committee will need to devote immediate attention to assessing available data, identifying its information needs, and developing plans for acquiring information needed to characterize air quality conditions, document sources of pollution, and compare the effectiveness of pollution control strategies. The following activities should be undertaken:

- *Expansion and integration of air quality monitoring network.* Reliable and regular information on air quality conditions throughout the international air basin is needed to understand the nature, extent, and magnitude of pollution problems. Monitoring also is needed to identify the sources of pollution problems. The Joint Committee will need to evaluate the adequacy of the existing air quality monitoring network; recommend additional sites or types of monitoring, including methods for tying together the respective governments monitoring efforts; and insure daily public access to monitoring data.
- *Completion of comprehensive emissions inventory.* The Joint Committee will need to examine existing emissions inventories and identify inventory needs. Although existing information may be adequate to begin developing pollution control and prevention strategies for known pollution sources, a comprehensive emissions inventory is needed to guide long-term planning. The Joint Committee will need to assert strong leadership to insure that data are collected as quickly as possible.
- *Development of analytical tools.* Air quality models will be needed to assess the effectiveness of various strategies. The committee should obtain information about current state-of-the-art models, including data needed to drive them and associated uncertainties, so it can make reasonable decisions regarding their use and the highest priority data needs. The Joint Committee should seek expert advice on how existing models could be refined or adapted.
- *Transferring technology and skills.* The Joint Committee could play a key role in continuing and expanding technology exchanges and technical training programs, so that local entities will have the technical hardware and expertise to operate and maintain their air monitoring networks and emissions inventories at comparable levels of efficiency.

Although a substantial base of data already exists to provide a basis for initial emission reduction plans, acquiring comprehensive air quality and emissions information will require significant resources and time. The Joint Committee should take a leadership role in obtaining the necessary commitments and financial resources from government and private interests. In the interim, actions can and should be taken, based on what is already known about the air pollution problem.

Pollution Abatement Projects

In order to demonstrate results in the short term, the Joint Committee should develop a list of pollution abatement projects to address known problems and strive to insure implementation of a few specific projects within its first year of operation. As a starting point, projects identified by the Paso del Norte Air Quality Task Force could be reviewed and supported or supplemented, if appropriate. In particular, the following projects merit attention:

- Continuing and expanded training programs for automobile mechanics.
- A mobile source management plan for the air basin, including investment in public transit and cleaner fuels.
- Workshops for paint and solvent shop businesses, including methods for reducing emissions through process changes (e.g., spray-painting booths) and material substitution (e.g., use of less reactive solvents).
- Brick oven conversions to natural gas.
- Installation of vapor recovery equipment at gas stations.
- Industry-sponsored automobile tune-ups for employees.
- Increased transborder sale of electricity and cleaner fuels.
- Use of waste materials to pave roads.
- Increased staffing and operational changes at international bridge crossings to reduce traffic congestion.

Most of these projects could be implemented through local cooperative efforts between the business community and state and local government agencies. An initial emphasis on pollution abatement projects that do not require federal government approval or action will help achieve short-term results and create a favorable track record for more complex projects.

Use of Economic Incentives

In selecting projects for implementation, the Joint Committee should place a high priority on cost-effectiveness and promoting transboundary investment in pollution control technology and strategies. The international agreement gives the Joint Committee a specific role in developing and implementing economic instruments, including studying the feasibility of a transboundary emissions trading program. The emphasis on economic instruments is consistent with the shift that is occurring in environmental policy generally in the U.S. and the specific encouragement of market-based programs in the federal Clean Air Act. Market-based programs promote innovation and the most cost-effective solutions to air pollution, by allowing industry greater flexibility and rewarding efforts that result in greater emission reductions than otherwise are required by law or regulations.[33]

There are several ways in which economic instruments could be used in the Paso del Norte transboundary air basin.[34] One of the most promising and intriguing options is to allow U.S. companies to earn emission reduction credits under the U.S. Clean Air Act for pollution control projects in Ciudad Juárez. The U.S. Clean Air Act requires that new industries wishing to locate in the El Paso nonattainment area must obtain emission reductions elsewhere in the area to offset pollution increases from the proposed facility. Furthermore, EPA and TNRCC have developed regulations for emission reduction credits, offsets, banking, and Area Emissions Reduction Credit Organizations.[35]

Many cost-effective emission reductions opportunities exist in Ciudad Juárez. U.S. companies would have a powerful incentive to seek out and implement the most cost-effective "offsets," if they could earn credit for pollution control projects in Ciudad Juárez. The following examples illustrate the type of opportunities that exist:

- A new El Paso–based manufacturer of electric motors emitting two hundred tons of volatile organic compounds per year might opt to meet its offset requirement by investing in Stage I vapor recovery systems in Ciudad Juárez rather than in a more costly alternative on the U.S. side of the border.
- A new or expanding cement plant could choose to offset increasing particulate matter emissions by agreeing to pave roads in Ciudad Juárez.
- A U.S.-based gasoline distributor could sell cleaner fuel in Ciudad Juárez or implement programs to retire high-emitting vehicles to offset emission increases associated with higher sales.
- Local government agencies could earn credit toward their annual emission reduction requirements by convincing federal authorities to modify operations of the international bridges to reduce emissions from vehicle idling.

- Electric utilities could earn credit to offset emissions resulting from construction or expansion of generating plants by selling electricity or investing in renewable energy development in Ciudad Juárez to replace existing high-polluting diesel generators.

To date, there has been one government-approved transboundary investment in the Paso del Norte air basin. It was completed in April, 1997, when El Paso's GI Corporation agreed to satisfy an EPA penalty by investing in new pollution-reducing soldering machinery for its maquiladora facility in Ciudad Juárez.[36] This agreement has provided the U.S. and Mexico with valuable experience in transboundary environmental management. The Joint Committee now is working with the governments to develop a list of future projects to improve air quality in the region.

The EDF and others have begun work on guidelines for a transboundary emission reduction credit program that could be refined and adapted for use. The paramount concern is insuring that reductions achieved comply with the fundamental requirements of EPA's emissions reduction credit program. The Paso del Norte Air Quality Task Force already is working to help identify industries seeking to expand or locate in El Paso which might be good candidates for demonstration projects. Identifying and conducting prototype projects not only would achieve pollution reductions but also would help solve many of the legal, administrative, political, and scientific problems associated with transboundary emissions trading. In particular, in order to insure the credibility and integrity of an emission reduction credit program, mechanisms must be developed (and possibly tailored to each case) to insure that reductions achieved meet the following criteria:

1. *Quantifiable.* Develop pre- and post-construction/implementation requirements for monitoring, reporting, and record keeping. These requirements must outline how to quantify emission reductions obtained and demonstrate air quality improvement.
2. *Surplus.* Develop procedures for defining the baseline emissions level from which emission reductions may be credited, using each country's and state's existing laws and regulations as a benchmark. For example, the following types of activities may produce surplus reductions:
    a. A permanent shutdown of a facility, coupled with a demonstration that emission reductions will not be negated by construction of a similar emission-producing activity.
    b. Installation of a level of control which is greater than

required by regulations or state implementation plan provisions.
c. Installation of different processes or equipment which emit less pollutants than previous processes or equipment.
d. More effective operation and maintenance of abatement and process equipment.
e. Verifiable and permanent reduction in production rates or hours of operations.
f. Utilization of alternative fuel vehicles beyond what is required by law.

3. *Permanent.* Develop procedures for demonstrating that emission reductions will be contemporaneous with, and assured for, the life of the corresponding increase. This may include requirements for participation in training programs to acquire skills needed continuously to maintain and monitor the efficiency of pollution-control equipment or process changes.

4. *Enforceable.* Develop requirements that emission reduction strategies be set forth in permit conditions or regulation changes, memoranda of understanding between government agencies, or contracts between private parties. Legal enforceability must be coupled with an assurance of practical enforceability—i.e., resources must be available to devote to inspection and enforcement activities. Emission fees or transaction fees could be established for sources wishing to engage in the transboundary emission reduction credit program, with the fees being used to augment existing enforcement resources.

Alternatively, if a contractual arrangement is used, the contract between affected parties could incorporate certain incentives to insure the permanence and enforceability of the emission reductions. For example, if a U.S. firm has financed the cost of pollution-control equipment and periodic loan payments are established, the payment schedule could be accelerated in the event of noncompliance; or the sponsoring company could agree to pay its partner periodically for operation and maintenance costs so long as compliance occurs. The contract also could include a liquidated damages and arbitration provision to avoid potential litigation. Finally, the contract could designate the Joint Committee as a third-party beneficiary, so that independent enforcement action could be taken by governmental entities in the event of noncompliance.

As mentioned above, specific requirements likely will need to be developed on a case-by-case basis, as transboundary emission reduction opportunities arise. The Joint Committee should conduct an outreach program to help solicit companies to participate in demonstration projects, so that this promising approach to reducing pollution can be tested and implemented more broadly in the air basin. The Joint Committee could serve as the "certifying" agency, by helping to negotiate appropriate conditions, develop suitable instruments, review and verify the validity of emission reduction credit transactions, and monitor compliance. In addition, any application for obtaining transboundary emission reduction credits would need to be subjected to public review, and credit could not be awarded until enforceable mechanisms are in place.

Public Education

The Joint Committee should have a central role in educating the public about air pollution problems and disseminating information about how individuals can reduce pollution through their own actions and behavior. To increase public awareness, the Joint Committee's work plan should include preparing regular (at least annual) reports to the public on air quality conditions and trends throughout the international air basin. Once monitoring networks are integrated and automated, daily reports on air quality conditions should be issued. Publicizing air quality conditions on a daily basis, along with information about associated health effects and measures that can be taken to reduce pollution, promotes public awareness. The Joint Committee also may sponsor workshops and prepare informational materials for businesses and schools outlining pollution prevention and reduction strategies.

Public education efforts should be used to demonstrate accountability of the Joint Committee. The annual report should summarize recommendations made by the Joint Committee and action taken on the recommendations. Pollution abatement projects implemented should be described and their effectiveness assessed. The report also should disclose financial aspects of the Joint Committee's activities and projects, including an accounting of funding received and how it was used. Through these reports, the public will gain confidence that the Joint Committee is an effective and efficient institution—not just another layer of bureaucracy.

CONCLUSIONS

The agreement signed by the U.S. and Mexican governments on May 7, 1996, represents a step toward better regional management of transboundary air pollution. This result came about through the effective work of the Paso del Norte Air Quality Task Force, which put forth a strong proposal and got most of what it asked for. Although the federal governments necessarily remain major players with important authorizing and oversight roles, the international air quality management basin is a reality, and the Joint Committee will promote greater community involvement in decision making and pollution abatement activities.

In particular, the Joint Committee's members will be representative of the local binational community and the committee will have broad authority to: establish its own operating and administrative procedures; engage in technology transfer and training programs; generate and disseminate information for the public; serve as a forum for hearing public concerns about air pollution problems and potential solutions; conduct joint studies and analyses of air pollution problems, including the integration and automation of monitoring networks; and develop, and in many cases implement, pollution control and prevention projects. The Joint Committee also will be accountable to the public and will serve to hold the federal governments accountable for acting upon recommendations. Perhaps one of the Joint Committee's most important functions will be to marshal the financial and human resources needed to bring about real air quality improvement, including economic incentives for transboundary investment in pollution control.

As an unprecedented, evolutionary, and community-led effort toward institutionalizing shared responsibility for managing transboundary air pollution problems, the progress made by the Joint Committee will need to be carefully monitored and periodically evaluated. As barriers are identified, creative approaches to overcoming them can be devised. Perhaps the greatest potential lies in carrying out pollution reduction projects that help achieve cleaner air and support economic growth, without getting bogged down in conflicts that arise because of the international border or differing stages of economic development. At some point in the future, it may be helpful to develop a time schedule and procedure for moving from transboundary governance based on differing domestic standards to a set of uniform and consistent ambient air quality standards, coupled with the establishment of emission budgets for the various pollutants and source sectors that contribute to air pollution problems in the international air basin.

Most importantly, the Joint Committee and the Paso del Norte Air Quality Task Force will help the citizens of the region take responsibility for achieving cleaner air. With strong community support, the Joint Committee can implement a transboundary air quality management program that promotes cooperation among the parties, minimizes the cost of pollution control, and delivers better air quality. If the new governance structure proves to be effective, it will serve as a good model for other border communities struggling with transboundary air pollution problems or other shared environmental problems. Citing the Paso del Norte experience, leaders in San Diego and Tijuana now are investigating the possibility of a transboundary air quality management program for their region. In addition, a recent agreement between the U.S. and Canada to reduce crossborder transport of ozone and particulate matter contains provisions similar to those in the Paso del Norte accords.[37]

APPENDIX A
Annex VI Proposed by the Paso del Norte Air Quality Task Force

ANNEX VI TO THE AGREEMENT
BETWEEN
THE GOVERNMENT OF THE UNITED STATES OF AMERICA
AND
THE GOVERNMENT OF THE UNITED MEXICAN STATES
ON COOPERATION FOR THE PROTECTION AND IMPROVEMENT
OF THE ENVIRONMENT IN THE BORDER AREA

AGREEMENT OF COOPERATION
BETWEEN
THE GOVERNMENT OF THE UNITED STATES OF AMERICA
AND
THE GOVERNMENT OF THE UNITED MEXICAN STATES
REGARDING THE FORMATION AND PURPOSES OF AN
INTERNATIONAL AIR QUALITY MANAGEMENT DISTRICT

The Government of the United States of America ("the United States") and the Government of the United Mexican States ("Mexico") ("the Parties"),

Recognizing that substantial improvement of air quality is needed in their common border zone;

Realizing that such air quality improvement can occur most effectively and efficiently with increased cooperation;

Recognizing that the formation of a joint district to work for such air quality improvement is desirable in the El Paso, Texas—Ciudad Juarez, Chihuahua—Dona Ana County, New Mexico area;

Reaffirming Principle 21 of the 1972 Declaration of the United Nations Conference on the Human Environment, adopted at Stockholm, which provides the Nations have, in accordance with the Charter of the United Nations and the principles of international law, the sovereign right to exploit their own resources pursuant to their own environmental policies and the responsibility to ensure that activities within their jurisdiction or control do not cause damage to the environment of other Nations or areas beyond the limits of national jurisdiction;

Recognizing that Article 3 of the Agreement between the Parties on Cooperation for the Protection and Improvement of the Environment in the Border Area of 1983 ("the 1983 Agreement") provides that the parties may conclude specific arrangements for the solution of common problems in the border areas as annexes to that Agreement,

Have agreed as follows:

ARTICLE I
GENERAL PURPOSE

In cooperation with existing National, State, and Local authorities, the Parties agree to hereby establish an International Air Quality Management District ("IAQMD") on their common border.

ARTICLE II
APPLICABILITY

1. The IAQMD shall encompass the following geographic area of jurisdiction: El Paso County, Texas; Dona Ana County, New Mexico; and the greater Metropolitan Area of Ciudad Juarez, Chihuahua.

2. Nothing in this agreement will supersede the obligations of affected United States State and Local authorities or affected Mexico State and Local authorities to meet all existing applicable rules and regulations of the United States and Mexico, respectively.

ARTICLE III
SPECIFIC PURPOSES

The purposes of the IAQMD will be as follows:

(1) to develop a consolidated air emissions inventory and ambient air quality data base

and, using environmental targets established by the Parties, including those based on health considerations,

(2) To develop options for both short-term and long-term air pollution control and prevention measure, including an emergency action plan

(3) to identify optimum, joint control strategies, including an economic incentives-based program to encourage investment in pollution control projects and similar innovative mechanisms of pollution control

(4) to recommend joint control strategies to the respective National Governments

(5) to work with the respective National, State, and Local Governments to implement joint control strategies and to promote transboundary transfer of technology and skills to facilitate the optimization of such strategies

(6) to educate the general public regarding air quality control issues

(7) to establish procedures to allow effective and efficient communication among the parties

(8) to establish cooperative compliance efforts between the respective governments, provide a forum for citizen participation, and jointly develop recommendations for appropriate remedial actions by the respective governments.

ARTICLE IV
STRUCTURE OF GOVERNING BOARD

The Parties agree to establish a Governing Board ("the Board") for the IAQMD. The Board shall be established according to the following guidelines:

(1) The Board shall be appointed by the Parties. The Board shall consist of representatives from the Federal, State (Texas, New Mexico, Chihuahua), and Local (El Paso, Ciudad Juarez, Dona Ana County) Governments. The Parties shall each appoint two members-at-large representing the public, including at least one each representing local industry. All Board members shall serve a three-year term.

(2) The Board shall have the authority to appoint an Executive Director and technical staff to enable it to carry out its responsibilities. The size and composition of the staff shall be determined by the Board, consistent with operational needs.

(3) The Parties agree to provide sufficient funding and personnel to ensure the IAQMD's functions can be adequately carried out.

ARTICLE V
RELEASE OF INFORMATION TO THIRD PARTIES

The Parties shall follow the guidelines set forth in Article 16 of the 1983 Agreement related to the procedure for sharing technical information with third parties and Article VI of Annex V of the 1983 Agreement for establishing procedures to protect the confidentiality of proprietary or sensitive information conveyed between the Parties pursuant to this Annex.

ARTICLE VI
EFFECT ON OTHER INSTRUMENTS

Nothing in this Annex or its appendices shall be construed to prejudice other existing or future agreements concluded between the Parties or affect the rights or obligations of the Parties under international agreements to which they are party.

ARTICLE VII
IMPLEMENTATION

Implementation of this Annex is dependent upon the availability of sufficient funding.

ARTICLE VIII
APPENDICES

Appendices to this Annex may be added through an exchange of diplomatic notes and shall form an integral part of this Annex.

ARTICLE IX
AMENDMENT

This Annex, and any appendices added thereto, may be amended by mutual agreement of the Parties through an exchange of diplomatic notes.

ARTICLE X
REVIEW

The National Coordinators under the 1983 Agreement or their designees shall meet at least every year from the date of entry into force of this

Annex, at a time and place to be mutually agreed upon, in order to review the effectiveness of its implementation and to agree on whatever individual and joint measures are necessary to improve such effectiveness.

ARTICLE XI
ENTRY INTO FORCE

This Annex shall enter into force after signature when each Party has informed the other through diplomatic notes that it has completed the internal procedures necessary for the Annex to enter into force.

ARTICLE XII
TERMINATION

This Annex shall remain in force indefinitely, unless one of the Parties notifies the other in writing through diplomatic channels of its desire to terminate it, in which case the Annex shall terminate six months after the date of such written notification.

IN WITNESS WHEREOF the undersigned, being duly authorized by their respective Governments, have signed this Annex.

Done at __(location)__, in duplicate, this __(date)__ in the English and Spanish languages, both texts being equally authentic.

FOR THE GOVERNMENT OF THE
UNITED STATES OF AMERICA:

FOR THE GOVERNMENT OF THE
UNITED MEXICAN STATES:

APPENDIX B

Agreement to Establish a Joint Advisory Committee for the Improvement of Air Quality

APPENDIX I

ANNEX V TO THE AGREEMENT BETWEEN THE
UNITED MEXICAN STATES
AND THE UNITED STATES OF AMERICA ON THE
COOPERATION FOR THE
PROTECTION AND IMPROVEMENT OF THE ENVIRONMENT
IN THE BORDER AREA

AGREEMENT OF COOPERATION BETWEEN THE
UNITED MEXICAN STATES
AND THE UNITED STATES OF AMERICA REGARDING
INTERNATIONAL TRANSPORT OF URBAN AIR POLLUTION

Recalling that in the preamble to Annex V the Parties affirm their intention to ensure a reduction in air pollution concentrations for the benefit of their citizens living in the urban areas along the United States-Mexico border; and

Recognizing the importance of the participation of the local communities in carrying out the efforts to achieve this objective;

The Parties, having decided to establish a Joint Advisory Committee for the Improvement of Air Quality (hereinafter "the Committee") in the Ciudad Juarez, Chihuahua/El Paso, Texas/Dona Ana County, New Mexico air basin (hereinafter "air basin"),

Have agreed as follows:

DEFINITION
The air basin is defined as the geographic area that includes El Paso County, Texas and those parts of Dona Ana County, New Mexico and the metropolitan area of Ciudad Juarez, Chihuahua that are within 100 km. of the border.

OBJECTIVE

The Committee is established for the purpose of developing and presenting recommendations to the Air Work Group established under the La Paz Agreement regarding strategies for the prevention and control of air pollution in the air basin.

SCOPE OF ACTIVITIES

The Committee may develop recommendations for the Air Work Group on:

a) The joint development of studies and analyses on air quality monitoring and modeling, and air pollution prevention and abatement strategies in the air basin;

b) Exchanges of information on air quality matters such as air quality data, air emissions data, and data on compliance with each Party's air standards;

c) Technical assistance programs, technology exchanges, and training in areas relevant to preventing and reducing air pollution in the air basin;

d) Environmental education and outreach programs for the general public relevant to preventing and reducing air pollution in the air basin;

e) Exploring strategies to prevent and reduce air pollution in the air basin, including recommendations on emissions trading and other economic incentives as well as improving the compatibility of air quality programs in the air basin; and

f) Such other air quality improvement issues as the Committee may deem to be pertinent to the air basin and as may be recommended by the Parties.

The Parties will provide a guidance document to the Committee detailing more specific subject areas which the Committee should consider. This guidance document may be updated periodically by the Parties.

The recommendations may include analyses of the estimated costs, and possible financial sources, to implement the recommendations. The recommendations may also address the availability of technology and training necessary for their implementation.

STRUCTURE AND ORGANIZATION

The Committee will consist of 20 persons, ten of whom are to be selected by each Party, in close consultation with state and local governmental officials and the public in the air basin.

The ten U.S. representatives invited to serve on the Committee will include (i) one representative of the federal government; (ii) one representative from each of the governments of the States of Texas and New Mexico; (iii) one representative from local government in El Paso, Texas; (iv) one representative from local government in Dona Ana County, New Mexico; and (v) five persons, residing in the air basin, who are not employed by federal or any state or local government. At least one of these five persons will be a representative of the business community and at least one will be a representative of a nongovernmental organization, a major portion of whose activities concerns air pollution.

The ten Mexican representatives invited to serve on the Committee will include (i) one representative of the National Institute of Ecology (INE-SEMARNAP); (ii) one representative of the Federal Attorney for Environmental Protection (PROFEPA); (iii) one representative of the Federal health and Welfare Agency (SSA); (iv) one representative of the environmental authorities of the State of Chihuahua; (v) one representative of the environmental authorities of the Municipality of Ciudad Juarez; and (vi) five Mexican citizens, residing in Ciudad Juarez, who are not employed by federal, state, or local government. At least one of these five persons will be a representative of the private sector, at least one will be a representative of a nongovernmental organization, a major portion of whose activities concerns air pollution, at least one will be a representative of the academic institutions of Ciudad Juarez, and at least one will be a representative of the Consulting Council for Sustainable Development in the Northern Region.

One federal representative from each side will preside over the Committee. The Committee will make decisions by consensus.

The Committee will establish its own rules of procedure, subject to approval by the Parties. Meetings of the Committee will generally be open to the public.

The Air Work Group will consider the recommendations of the Committee and inform the Committee of any action taken pursuant to such recommendations.

The recommendations of the Committee will not be binding on the Air Work Group or the Parties.

REVIEW AND TERMINATION

The Parties will periodically review the implementation of this Appendix.

This Appendix will remain in force indefinitely, unless one of the Parties notifies the other in writing through diplomatic channels of its intention to terminate it or Annex V, in which case the Appendix shall terminate six months after the date of such notification.

NOTES

1. City of El Paso, Texas, Department of Planning, *El Paso Metropolitan Plan,* 1996. Ciudad Juárez, Chihuahua, Planning Department, *Plan Director de Desarrollo Urbano,* 1995.
2. EPA, "Compendium of EPA Binational and Domestic U.S.-Mexico Activities," EPA 160-B-95-001 (Washington, D.C.: GPO, 1995).
3. Laura Margarita Uribarri, "The Paso del Norte Air Quality Task Force: A Case Study in Binational Cooperation," senior honors thesis, International Relations, Stanford Univ., 1996.
4. EPA, Region 6, "An Overview of U.S.-Mexico Border Air Activities Schedule Through the Integrated Border Environmental Plan," by James W. Yarbrough, unpublished report, Document No. 92-165.07, Dallas, TX, 1996.
5. John C. Ross, Jr., "Do the Existing Air Pollution Statutes and Regulations Provide the Required Protection in the El Paso–Juárez Area?" in *Air Pollution Along the U.S.-Mexico Border,* ed. Howard G. Applegate and C. Richard Bath (El Paso: Texas Western Press, UTEP, 1974), 115–19.
6. EPA and Mexico, Secretaría de Desarrollo Urbano y Ecología, "Integrated Environmental Plan for the Mexican-U.S. Border Area," EPA A92-171 (Washington, D.C.: EPA, 1992).
7. Ibid.
8. Larry Butts, *Air Quality Assessment Program* (Austin: Texas Natural Resource Conservation Commission, 1996).
9. Maria M. Aponte-Pons and Alison A. Miller, "Air Pollution and the Texas-Mexico Border: Recent Research Initiatives," paper presented at 87th Annual Meeting, Air and Waste Management Association, Cincinnati, Ohio, 1994.
10. Butts, *Air Quality Assessment.*
11. W. Einfeld and H. W. Church, "Winter Season Air Pollution in El Paso–Ciudad Juárez: A Review of Air Pollution Studies in an International Airshed," SAND-95-0273 (Albuquerque, N.Mex.: Sandia National Laboratories, 1995).
12. Archie Clouse, *El Paso Air Quality Fact Sheet* (El Paso: Texas Natural Resource Conservation Commission, 1995).

13. Einfeld and Church, "Winter Season Air Pollution."
14. Howard G. Applegate, C. Richard Bath, and Jeffrey T. Brannon, "Binational Emissions Trading in an International Airshed: The Case of El Paso and Ciudad Juárez," *Journal of Borderland Studies* (Las Cruces: New Mexico State Univ.) 4, no. 2 (1989): 1–25.
15. Clouse, *El Paso Fact Sheet.*
16. Christopher J. Kennedy, "Innovations in Binational Management of the Border Environment: The Case of Air Quality in El Paso del Norte," paper presented at 87th Annual Meeting, Air and Waste Management Association, Cincinnati, Ohio, 1994.
17. Einfeld and Church, "Winter Season Air Pollution."
18. C. Richard Bath, "Air Pollution Regulation and the Question of Environmental Equity: A Case Study of El Paso, Texas, Ciudad Juárez, Chihuahua, and Sunland Park, New Mexico," unpublished report, Univ. of Texas at El Paso, 1994.
19. City of El Paso, *El Paso Metropolitan Planning.*
20. General Law of Ecological Equilibrium and the Protection of the Environment, *Diario Oficial,* art. 3, para. 11 (Jan. 28, 1988).
21. See U.S. Clean Air Act, 42 USC 7401-7671; EPA, *State Implementation Plans: General Preamble for Implementation of Title I of the Clean Air Act Amendments of 1990,* 57 Fed. Reg. 13498 (Washington, D.C.: GPO, 1992).
22. EPA and Mexico, Secretaría de Desarrollo Urbano y Ecología, "Integrated Environmental Plan."
23. EPA, "U.S.-Mexico Border XXI Program," EPA 160-D-96-001 (Washington, D.C.: EPA, 1996).
24. Ross, "Do the Existing Air Pollution Statutes?"
25. Francis S. Ainsa, Jr., "International Cooperation in the Abatement of Environmental Air Pollution in the El Paso–Juárez Area," in Applegate and Bath, *Air Pollution Along the U.S.-Mexico Border,* 125–30.
26. César Sepúlveda, "Métodos intergubernamentales viables para la cooperación en el control y eliminación de la contaminación del aire a lo largo de la frontera Mexico-Norteamericana," in Applegate and Bath, *Air Pollution Along the U.S. Border,* 131–36.
27. Applegate, Bath, and Brannon, "Binational Emissions Trading."
28. Ralph K. M. Haurwitz, "U.S., Mexico Agreement Offers a Breath of Fresh Air," *Austin (Tex.) American-Statesman,* Apr. 17, 1996.
29. Andy Pasztor, "U.S. Mexican Officials Plan to Create Air Pollution Zone for Border Residents," *Wall Street Journal,* 1996.
30. Kirk P. Watson and Peter M. Emerson, "Border Towns Need Air Pollution Program," *Austin (Tex.) American-Statesman,* July 14, 1993.
31. Allen Blackman and Geoffrey J. Bannister, "Crossborder Environmental Management and the Informal Sector: The Ciudad Juárez Brickmakers' Project," this volume.
32. Warren Christopher, U.S. Department of State, "American Diplomacy and the Global Environmental Challenges of the 21st Century," speech at Stanford University, Stanford, Calif., Apr. 9, 1996.

33. Daniel Dudek and John Palmisano, "Emissions Trading: Why Is This Thoroughbred Hobbled?" *Columbia Journal of Environmental Law* 13 (1988): 217–56.
34. Kathryn C. Wilson, "The International Air Quality Management District: Is Emissions Trading the Innovative Solution to the Transboundary Air Pollution Problem?" *Texas International Law Journal* 30, no. 2 (1995): 369–93.
35. EDF, *Emissions Trading in Nonattainment Areas* (New York: EDF, 1995).
36. EPA, Region 6, "In the Matter of GI Corporation, Hatboro, Pennsylvania," Consent Agreement and Consent Order, Docket Number RCRA VI-328-H, Dallas, Tex., Apr., 1997.
37. "Program to Develop Joint Plan of Action for Addressing Transboundary Air Pollution," signed by Carol Browner, administrator, EPA, and Hon. Sergio Marchi, minister of the environment, Mexico, Apr. 7, 1997.

6

CROSSBORDER ENVIRONMENTAL MANAGEMENT AND THE INFORMAL SECTOR
CIUDAD JUÁREZ BRICKMAKERS' PROJECT

ALLEN BLACKMAN
AND GEOFFREY J. BANNISTER

The considerable difficulties associated with crossborder environmental management are compounded when polluters are located in the informal sector—that is, the sector comprised of low-technology microenterprises such as auto repair shops, finishing shops, and traditional brick kilns that operate outside the purview of any regulatory or tax authority.[1] Informal sector firms (which account for over one-half of nonagricultural employment in most Latin American and African countries, including Mexico), are difficult to regulate for four reasons.[2] First, by definition, they have few preexisting ties to the state. Second, they are small, numerous, and geographically dispersed. Third, they generally operate in highly competitive markets and, as a result, are under considerable pressure to cut costs, regardless of the environmental impacts. Fourth, they sustain the poorest of the poor and therefore may appear to both regulators and the public to be less appropriate targets for regulation than larger, wealthier firms.

Given these constraints, even if crossborder coordination between environmental authorities can be achieved, conventional command-and-control regulation is unlikely to work. New approaches are called for. The Ciudad Juárez Brickmakers' Project, which has worked to abate highly polluting emissions from the city's approximately 350 small-scale brick kilns since 1990, is an example of such an approach. This project is innovative, first, because it is headed not by an environmental authority but by a nongovernmental organization, the Mexican Federation of Private Health Associations and Community Development (FEMAP, for its initials in Spanish). In addition, it has:

- Enlisted the participation of private and public sector stakeholders from both sides of the border.
- Focused on developing and disseminating pollution prevention technologies.
- Worked to establish a cooperative, rather than an adversarial, relationship with brickmakers, creating incentives for voluntary compliance as well as penalties for noncompliance.

This chapter reviews the history of the Brickmakers' Project in order to distill lessons about crossborder environmental management in the informal sector. We draw upon a variety of sources, including an extensive survey (administered to ninety traditional brickmakers in July, 1995), interviews with project participants, and primary and secondary written sources.[3]

The Brickmakers' Project did not achieve its immediate objective of converting the brickmakers in Ciudad Juárez to clean-burning propane—as of current writing, virtually all of them have reverted to burning debris. Nevertheless, its innovative efforts to build a multisector crossborder coalition and to create incentives for pollution prevention in the informal sector were successful and probably are harbingers of things to come.

We draw three lessons from the project's history. First, private-sector-led crossborder initiatives can work—indeed, they may be more effective than public-sector initiatives—but they require strong public-sector support and some ability on the part of project organizers to leverage this support. Second, necessary conditions for effective environmental management in the informal sector include: enlisting the cooperation of local unions and political organizations, relying upon peer monitoring among informal firms, and providing inducements to offset compliance costs. Ineffective strategies include: promoting too-advanced and therefore inappropriate technologies, and intervening in informal markets. Finally, the history of the Brickmakers' Project suggests that, in volatile developing economies, even well-designed voluntary market-based environmental initiatives in the informal sector are bound to be fragile.

The next section of this chapter provides background on air pollution in El Paso–Ciudad Juárez and on traditional brickmaking. The third section details the history of the Brickmakers' Project, and the concluding section distills lessons from this history.

BACKGROUND

El Paso–Ciudad Juárez Air Quality

Air quality in the sister cites of El Paso, Texas, and Ciudad Juárez, Chihuahua, is the worst on the U.S.-Mexican border and among the worst in North America.[4] The problem stems from both the geography of the region (which fosters temperature inversion) and from pollution created by rapid industrialization and population growth over the last several decades. From 1980 to 1990, the population of Ciudad Juárez grew at an average annual rate of 3.47 percent, while that of Mexico as a whole grew at 1.92 percent. During the same time period, the population of El Paso grew at an average annual rate of 2.79 percent, while that of the U.S. grew at an average annual rate of 1.26 percent.[5]

The leading sources of air pollution in the region, in order of magnitude, are: vehicle emissions, dust from unpaved roads, industrial pollution, and open-air fires. Not surprisingly, the locations of these sources reflect the relative levels of development on the two sides of the border. Open-air fires used in brickmaking and residential heating, unpaved roads, cement plants, and a relatively old vehicular fleet in Ciudad Juárez are major sources of particulate matter and carbon monoxide. North of the border, the ASARCO copper smelter and the Chevron oil refinery are major sources of sulphur oxides, nitrogen oxides, and heavy metals. As of this writing, the city of El Paso was classified by the U.S. Environmental Protection Agency (EPA) as a "moderate" nonattainment area for both carbon monoxide and particulate matter, and El Paso County was classified as a "serious" nonattainment area for ozone. Although monitoring and record keeping in Ciudad Juárez are too poor to know for certain, it is quite likely that air quality in that city is very similar to that in El Paso.[6]

Binational efforts to control air pollution in Ciudad Juárez–El Paso date back at least to the 1970s but thus far have not resulted in much concerted action. According to Richard Bath, key reasons on the U.S. side include lack of coordination among different levels of government, lack of public support, and overly ambitious abatement goals; and, on the Mexican side, a dearth of financing and commitment. Moreover, binational cooperation has been hampered by sovereignty concerns.[7] Recent binational efforts, discussed by Angulo, Emerson, and Rincón in their chapter of this book, have focused on the creation of an air quality management district in El Paso–Ciudad Juárez.

Traditional Brickmaking in Ciudad Juárez

Ciudad Juárez's approximately 350 small-scale brick kilns traditionally have been fired with inexpensive, highly polluting fuels such as garbage, used tires, and wood scrap (often impregnated with toxic resins, laminates, and varnishes).[8] As a result, a number of sources contend that brick kilns are the third or fourth leading contributor to air pollution in the Ciudad Juárez–El Paso area.[9] Although brick kilns are associated primarily with carbon-monoxide emissions, depending on the fuels used they also emit particulate matter, volatile organic compounds, nitrogen oxide, sulphur dioxide, heavy metals, and carbon dioxide, the most important greenhouse gas. According to Johnson et al., tests of emissions from traditional brick kilns burning five different fuels—sawdust, contaminated sawdust, used motor oil, propane (old burner), and propane (new burner)—showed the two "least desirable" fuels to be used motor oil and contaminated sawdust. These emitted relatively high levels of volatile organic compounds and carbon monoxide and contained high concentrations of hazardous metals likely to be released into the air upon burning.[10]

Ciudad Juárez's brick kilns are clustered in eight poor *colonias* (neighborhoods) located throughout the city: Anapra, División del Norte, Francisco Villa, Fronteriza Baja, Kilómetro 20, México 68, Satélite, and Senecu 2. When brickmakers squatted in these colonias twenty-five or thirty years ago, all were situated on the outskirts of the city. However, over time, urban sprawl has enveloped most of them, and traditional kilns now can be seen along main avenues and near the major industrial parks. Because of their central locations, these kilns have been a principal subject of complaints to the Ciudad Juárez municipal environmental authority. Brick kilns were the most frequent subject of complaints (one in every four) to the Ciudad Juárez municipal environmental authority in 1994.[11]

On average, each of Ciudad Juárez's 350 kilns employs six workers; as a result, brickmaking is a significant source of employment.[12] According to FEMAP, the nongovernmental organization that leads the Brickmakers' Project, brickmaking in Ciudad Juárez provides over 2,000 jobs directly, plus 150 jobs indirectly in transportation and wholesaling. Brickmaking in Ciudad Juárez is a microcosm of a national industry composed of an estimated 15,000 operations, supporting 100,000 families, and most traditional kilns are concentrated in or near large urban areas such as Saltillo, Guadalajara, Monterrey, and the Federal District. FEMAP estimates that 2,500 to 3,000 of these kilns are situated in border cities.[13]

Traditional brickmaking is an extremely labor-intensive activity. All required tasks are performed by hand. First, clay, earth, and water are mixed together, sometimes with a pinch of sawdust or other organic material. Often the earth and clay are extracted from the areas surrounding the kilns, resulting in a pitted landscape.[14] The mixture then is molded into bricks, which are dried in the sun for one to three days, depending on the weather. Each worker can make up to a thousand bricks in one twelve-hour day. Once dry, the bricks are loaded into an adobe kiln. The kiln is a square roofless structure that contains a number of brick arches built into a sunken floor. The arches both support the bricks stacked on top of them and form a combustion chamber underneath. Average kiln capacity is about 10,500 bricks (each 2 by 5 by 10.5 inches). After baking and cooling, bricks are unloaded from the kiln directly into a truck for transportation or onto a *patio* of land beside the kiln.

Each cycle of production (molding, drying, loading the kiln, firing, unloading) takes about eleven days on average, which theoretically allows for about 2.3 firings per month, or 27 firings per year.[15] In practice, the average number of firings is 1.5 per month, for two reasons. First, from November to March, rain and cold weather inhibit molding, drying, and firing. More important, in the last year, the market for bricks has been extremely soft, owing to a construction slowdown that has accompanied Mexico's macroeconomic downturn. In the summer of 1995, traditional brickmakers were holding extensive unsold inventories.

Studies have put the brickmakers' profit per firing (when burning debris) at anywhere from 300 to 600 pesos ($40 to $80 U.S., at an exchange rate of 7.5 pesos to the dollar).[16] This compares to the monthly minimum wage in northern Mexico of about 480 pesos ($64) and illustrates the very small margin of profit that brickmakers must work with.[17]

Most brickmakers are over forty years of age, quite old in comparison to the population as a whole. On average, they have five years of schooling, although many have much less, and approximately a quarter cannot read or write. Living conditions are primitive. Most brickmakers live next to their kilns in small houses with no drainage or running water. The large majority have no access to health services.[18]

Fifty-nine percent of the brickmakers we surveyed belonged to a local organization. There are two main rival political factions among the brickmakers. The first is comprised of organizations affiliated with the nationally dominant Partido Revolucionario Institucional (PRI), such as the Federation of Mexican Workers (CTM; Confederación de Trabajadores Mexicanos) and the National Federation of Citizens' Organizations (FNOC; Frente Nacional de Organizaciones Ciudadanas). Many of the brickmakers in these organizations also belong to the PRI-affiliated Brickmakers' Union (Sindicato de

Ladrilleros y Trabajadores de la Cal). The PRI affiliates tend to represent the relatively affluent brickmakers. Their leaders act as intermediaries between the brickmakers, the city government, and federal agencies.

Because of their ties to the political establishment, be it PRI or PAN (Partido Acción Nacional, the PRI's principal opposition party), these organizations have been successful in extracting concessions for their members. For example, the leader of one brickmakers' union mentioned that members of his union received subsidies on water bills, as well as permits for dredging a local canal for clay. Brickmaker organizations were one of the main instruments used by the city government (at that time PAN) and FEMAP to help convert brickmakers to the use of propane gas.

PRI affiliates dominate certain brickmaking colonias, such as Satélite; have a sizable proportion of others, such as México 68; and are completely absent from some of the poorest colonias, such as Anapra. The poorest colonias are dominated by the rival Committee for Popular Defense (CDP; Comité de Defensa Popular), which is linked to the national Worker's Party (PT; Partido del Trabajo). The CDP traditionally has been opposed to the political establishment, having been formed to fight the city government's attempts to push out squatters. Members of the CDP continuously have resisted attempts by the Ciudad Juárez government to regulate brickmaking activities, and in particular they resisted adopting propane.

HISTORY OF THE BRICKMAKERS' PROJECT

Birth of the Brickmakers' Project

At every stage, the Ciudad Juárez Brickmakers' Project has been shaped by national and international political trends, as well as by local concerns and efforts. On the level of national politics, the project was, broadly speaking, spawned in the late 1980s by the Mexican federal government's new emphasis on environmental protection. Heralded by the emergence of the environment as an important issue in the 1988 federal elections and confirmed by the passage that same year of a comprehensive new General Law for Ecological Equilibrium and Protection of the Environment (the "Environmental Law"), this new environmentalism was by no means mere rhetoric. In just three years between 1988 and 1991, Mexican federal expenditures on the environment increased from $95 million to $1.8 billion, the equivalent of 0.7 percent of Mexico's gross domestic product.[19]

On the local level, an important antecedent to the Brickmakers'

Project was a growing recognition during the 1980s that traditional brick kilns in Ciudad Juárez are an important source of air pollution in the sister cities. An activist Citizens' Environmental Advisory Committee to the El Paso City Council played a pivotal role in creating this awareness.[20]

The first effort to introduce propane into the brickyards in Ciudad Juárez was the campaign undertaken in 1989 by Dr. René Franco Barreno, then director of the Municipal Council for Ecology in Ciudad Juárez. To our knowledge, this was the first concerted effort to abate emissions from brick kilns in Ciudad Juárez. Dr. Franco's short-lived campaign succeeded in enlisting a number of organizations that later played a key role in the Brickmakers' Project, notably the local office of the Federal Ministry of Commerce and Industry (SECOFI) and the association of local propane retailers. In 1990, the Municipal Council for Ecology turned the project over to FEMAP, a well-established private nonprofit organization based in Ciudad Juárez. Founded in 1973 by an energetic community leader, Sra. Guadalupe de la Vega, FEMAP administers health care and microenterprise development projects in twenty-five Mexican states and has had considerable experience working in Ciudad Juárez's poor colonias.

FEMAP's first step, in 1991, was to administer a census of brickmakers in Ciudad Juárez, covering such issues as production, markets, and socioeconomic status. FEMAP also cultivated ties to local propane companies. These companies ultimately stepped up their participation in the project, permitting impoverished brickmakers to purchase propane on credit, providing training and equipment (tanks, vaporizers, and burners) and donating glazed brick (used to reconstruct kiln arches to enable them better to withstand the heat generated by burning propane).[21]

Despite the assistance of the propane companies, for a number of reasons—including the inevitable difficulties associated with convincing an initial group of producers to risk changing centuries-old methods, a lack of strong regulatory pressure until 1992, and the time-consuming nature of grassroots organizing—the project did not succeed in converting more than 30 percent of brickmakers to propane until the end of 1993. As late as December, 1992, by FEMAP's count, no more than 15 percent of the brickmakers in five principal brickmaking colonias in Ciudad Juárez (some fifty to sixty brickmakers) were using propane.[22] Yet, even as early as 1991, the Brickmakers' Project had begun to generate an extraordinary amount of publicity, institutional participation, and some outside funding. What factors were responsible?

NAFTA and Mexican Politics

Two national and international events in the early 1990s focused unprecedented attention on border environmental issues: the NAFTA debate and the Mexican midterm federal elections of 1991. Both the Bush and the Salinas administrations committed enormous political capital to NAFTA. To the surprise of both administrations, in the spring of 1991, during the U.S. Congress debate over granting the Bush administration "fast track" authority to negotiate the treaty, environmentalists organized a vociferous opposition, arguing that a free trade agreement would spur still more rapid and uncontrolled development along the border. They argued that, despite the high level of environmental protection ostensibly afforded by Mexico's 1988 Environmental Law, enforcement in Mexico was extremely lax; therefore, a free trade agreement would create incentives for "dirty" U.S. firms to relocate in Mexico.[23]

The NAFTA negotiators responded by promising to prepare a master plan to deal with border environmental problems. But when the Integrated Environmental Plan for the U.S.-Mexican Border (IBEP) was released in draft form in August, 1991, it attracted scorching criticism in eighteen hearings in a number of border cities (including in El Paso) set up to elicit citizen "participation." Far from allaying environmentalists' concerns, the IBEP fueled the debate about the border environment, which continued unabated throughout the NAFTA negotiations and the subsequent congressional battle over ratification.

A second factor contributing to the heightened attention devoted to border issues in 1991 was the Mexican mid-term federal election. As in the 1988 election, environmental issues received considerable attention. Pre-election polls indicated that 60 percent of Mexicans considered the environment to be a high priority.[24]

The upshot of these events was that, by the fall of 1991, for the first time ever, the border environment became an important issue for both the U.S. and Mexican federal governments simultaneously. Both countries stepped up enforcement along the border and initiated a number of environmental projects and programs. Financing to deal with long-neglected border environmental issues, chronically scarce before, suddenly became available. Also, in November, 1991, with World Bank support, Mexico set in motion a plan to decentralize and reform its environmental regulation; the states were encouraged to pass and enforce their own environmental legislation subject to a federal floor established by the 1988 Environmental Law.[25]

The NAFTA debate, combined with Mexico's genuine, albeit recent, emphasis on the environment, created a special opportunity for FEMAP. The Brickmakers' Project was tailor-made for the political purposes of the Salinas administration. Not only did it involve cleaning up the border, but it also embraced private-sector initiative and the modernization of traditional microenterprises, two hallmarks of the administration's ambitious economic reforms. As an established ENGO, FEMAP was well placed to take advantage of this commonality of interests. The Brickmakers' Project also greatly appealed to Americans who were looking to defuse environmental opposition to NAFTA. Carol Browner, director of EPA, made a high-profile visit to a FEMAP demonstration site in March, 1993.[26]

Mexican Federal Support

In 1991, FEMAP was able to obtain an 800,000-peso trust fund for the project directly from the office of President Salinas through Solidarity Enterprises (Empresas Solidaridad), the microenterprise development branch of the urban development program PRONASOL.[27] This 800,000-peso fund then was used to leverage an 8-million-peso line of credit from NAFIN, the Mexican federal development bank, which finances small business projects. All of these funds were earmarked for the exclusive use of brickmakers.[28]

The Salinas administration made quite a show of its support for the Brickmakers' Project. Luis Donaldo Colosio, then head of SEDESOL and later the PRI's presidential candidate, delivered the first installment of his agency's funds personally in December, 1992. Salinas himself made public appearances with FEMAP officials in Ciudad Juárez on three separate occasions in February, 1993, August, 1993, and October, 1994.[29] Thanks to this high level of support, the project was well launched by 1994, the last year of the Salinas administration, and already was being heralded by U.S. and Mexican officials as a model of binational cooperation.

Participation in the Brickmakers' Project

One of the most noteworthy achievements of the Brickmakers' Project was constructing a broad base of institutional support that cut across national and economic-sector boundaries. A 1994 FEMAP report lists the following organizations as participants:

Public sector:

- SECOFI (the Federal Ministry of Commerce and Industry)
- NAFIN (a federal economic development bank)
- SEDESOL (the federal environmental agency, 1992–95)
- Municipal and state governments

- Universidad Autónoma de Ciudad Juárez
- INFONAVIT (the federal workers' housing agency)
- Solidarity Enterprises (a federal microenterprise development program, before 1992 administered within PRONASOL, after 1992 within SEDESOL)

Private Sector:

- FEMAP
- Asociación Gilberto (a Mexican charitable organization)
- Grupo Peñoles (a large Mexican mining conglomerate)
- Ciudad Juárez propane companies
- CONCANACO (a national federation of chambers of commerce)
- CANACINTRA (a national federation of manufacturing industries)
- COPARMEX (a federation of big business owners)
- Economic Development of Ciudad Juárez (a local businessmen's organization)
- Instituto Tecnológico y de Estudios Superiores de Monterrey ("Monterrey Tech"), Ciudad Juárez Campus
- Construction companies in Ciudad Juárez
- Brickmakers

United States:

- El Paso Natural Gas Company
- Los Alamos National Laboratory
- Gas Research Institute (Chicago)
- University of Texas at El Paso
- Southwest Center for Environmental Research and Policy (SCERP) (a consortium of U.S. universities)

While project organizers' hard work was instrumental in securing the participation of all these organizations, the publicity created by NAFTA, a commonality of interests between the project and the various organizations involved, FEMAP's well-established community ties, and a palpable bandwagon effect contributed as well. What roles did the various organizations play?

The national and local business federations—CONCANACO, CANACINTRA, COPARMEX, and Economic Development of Ciudad Juárez—were instrumental in building bridges to brickmakers' associations and to the construction industry. By law, all companies in Mexico must belong to chambers of commerce. The brickmakers were linked to CANACINTRA through various local organizations. The construction industry was well represented in CONCANACO, as well as COPARMEX. Grupo Peñoles contributed engineering expertise in very early efforts to improve kiln efficiency and donated materials to be used in adapting kilns to propane.[30] INFONAVIT issued an order in August, 1993, decreeing that all its contractors must use "ecological bricks"—bricks fired with propane—instead of "dirty" bricks and brick substitutes such as cinderblock.[31]

While NAFIN, the federal economic development bank, and Solidarity Enterprises, the federal microenterprise development program, participated as funders, the key public-sector actor was the municipal government of Ciudad Juárez and, in particular, the city's environmental authority. As is discussed below, for about a year, this office strictly enforced a prohibition on the burning of debris. The Chihuahua state government's principal role seems to have been to provide administrative, hortatory, and political support to the city government.[32] Personnel from Monterrey Tech at Ciudad Juárez helped to develop technical and management training courses for brickmakers.

American organizations also played noteworthy and highly visible roles. Both El Paso Natural Gas and Los Alamos National Laboratory provided engineers to design and test improved kilns and burners. In addition, El Paso Natural Gas provided at least $100,000 towards project operating expenses.[33] Funded by EPA via the Southwest Center for Environmental Research and Policy, U.S. universities became involved in developing brickmaker training courses and conducting research on methods of improving kiln fuel efficiency.[34]

As was true of their Mexican counterparts, the institutional interests of these American organizations favored participation in the Brickmakers' Project. El Paso Natural Gas's contributions to the project generated favorable public relations. Los Alamos National Laboratories' participation complemented its efforts to reposition itself as a center of environmental research and technology transfer to the private sector. American universities were driven by the availability of significant funding for border environmental projects. It bears emphasizing, however, that community spirit also clearly was an important motive for these organizations, especially for certain employees. Several of the project engineers from El Paso Natural Gas and Los Alamos National Laboratory often worked nights and weekends without remuneration, under difficult conditions.[35] This certainly was true also of FEMAP personnel, many of whom volunteered their time.

As for the brickmakers themselves, as discussed in the next section, though almost a third said that improving the environment was their key motivation for switching to propane, almost all mentioned various inducements or sanctions.

Strategies

In attempting to induce the brickmakers to switch to propane, project leaders and participants were faced with five key obstacles. First, adopting propane required brickmakers to obtain and learn to use relatively expensive new equipment. Second, at the beginning of the project, regulatory and hortatory pressure to abate kiln emissions was, for the most part, extremely weak. Third, although the brickmakers themselves were most affected by kiln emissions, most did not perceive them to be particularly harmful. Fourth, the traditional brick kiln industry and the construction industry that buys from it are intensely competitive; as a result, there was considerable pressure for individual brickmakers to cut costs by burning dirty fuels and for individual construction firms to purchase the least expensive building materials. Finally, and most important, throughout the project's life the cost of propane was higher than the cost of debris. Moreover, this cost differential increased significantly over time. In the early 1990s, PEMEX, Mexico's state-run petroleum company, began gradually to eliminate long-standing subsidies on propane in the border region (in conjunction with the Salinas administration's economic liberalization program and to dampen a black market in the U.S. for subsidized Mexican propane). At the beginning of 1991, propane prices were approximately 0.24 pesos per liter. For the average brickmaker, this implied a cost of 37.02 pesos per thousand bricks fired with propane, versus 29.77 pesos per thousand bricks fired with scrapwood—a difference in per-brick energy cost of 29 percent. By July, 1995, the price of propane had risen to 0.71 pesos per liter, implying an average cost of 109.52 pesos per thousand bricks fired with propane, versus 41.78 pesos per thousand bricks fired with scrapwood—a difference in per-brick energy cost of 162 percent.[36]

Participants in the Brickmakers' Project promoted a broad range of initiatives designed to overcome each of these barriers, including: donating propane equipment, setting up technical extension services, disseminating information on the health impact of burning debris, formal and informal regulation, technological innovation, and market intervention.

Propane Equipment

The Brickmakers' Project was quite successful in creating access to propane equipment. In our sample of fifty-nine propane users, equipment was provided free of charge by propane companies in Ciudad Juárez in every case (although in two cases the brickmakers themselves eventually purchased their own equipment). Of the thirty-six brickmakers we interviewed who never used propane, not one said that a key reason was that the required equipment had not been available or affordable.

Technical Extension

A number of organizations and individuals provided training in the use of propane, including: Ciudad Juárez gas companies (who seem to have taken the lead early on), FEMAP, El Paso Natural Gas, and extension agents from Monterrey Tech at Ciudad Juárez. FEMAP has attempted to institutionalize its extension services by establishing "ECO-TEC," a center devoted to applied research on brickmaking and to training brickmakers in management and the use of clean technologies. The center was built in the summer and fall of 1993, using land donated by the municipal government of Ciudad Juárez and funds from Solidarity Enterprises, El Paso Natural Gas, and the FEMAP Foundation, a fund–raising arm of FEMAP based in El Paso. The ECO-TEC complex includes office space for administrators, classrooms and dormitories for brickmakers, several experimental kilns, and brickmaking facilities.[37] FEMAP officials envision ECO-TEC as a national center of brickmaker research and training. Although today ECO-TEC plays a key role in FEMAP's brickmaker program, it did not begin operation until 1994, after the adoption of propane in Ciudad Juárez already had been derailed by increases in propane prices. It is unclear from our survey how important other sources of technical extension were in facilitating the adoption of propane. On one hand, 67 percent of fifty-eight brickmakers who used propane cited the "provision of information" as having played some role in their decision to adopt, and 9 percent cited it as the key reason. On the other hand, 88 percent of all respondents claimed that no "technical assistance" was available to them.

Regulatory Pressure

Project participants have been successful intermittently at promoting propane use by ratcheting up formal and informal penalties associated with burning debris. Prior to 1992, burning debris was more or less tolerated by municipal authorities. This tolerance began to evaporate when Ciudad Juárez elected a new municipal president, Francisco Villarreal, in November, 1991. Partly as a result of the political climate discussed above, Villarreal and the director of his ecology office, Francisco Núñez, orchestrated a crackdown on brick kiln emissions. The use of "dirty" fuel to fire brick kilns was banned, though the definition of what constituted "dirty fuel" changed over time.[38] A peer monitoring mechanism was instituted to facilitate enforcement:

citizens were encouraged to call Núñez's office with complaints about brick kiln emissions. Núñez's office then dispatched an enforcement team which routinely jailed violators for twenty-four to thirty-six hours and sometimes fined them as well. For several months in late 1992 and 1993, propane was the only permissible clean fuel. (While enforcement during this period was relatively vigorous, it never was universally effective. A significant proportion of brickmakers, at least 30 percent, continued to burn debris throughout the strict enforcement regime).

FEMAP administrators were supportive of this crackdown. It did not last, however. As propane prices continued to rise in 1993, opposition to the new regime grew among brickmakers. Eventually the municipal government relaxed the rule, officially permitting brickmakers to burn sawdust untainted by resins or varnishes. Recently this regulatory scheme has been dismantled. Beginning in 1995, PROFEPA, the Federal Attorney General's Office for Environmental Protection (created in 1992), assumed primary responsibility for enforcement along the U.S.-Mexican border.[39] The regulation of brick kiln emissions has slackened relative to the Villarreal regime.[40]

Informal regulation, actively encouraged by project organizers, also influenced brickmakers' production decisions. Both FEMAP and the city authorities worked intensively with leaders of the local brickmakers' organizations to encourage propane use. In March, 1993, the leaders of all of the brickmakers' organizations were brought together to hammer out an agreement on permissible fuels and to set a deadline for the switch to such fuels.[41] Some of the brickmaker organizations were quite cooperative, especially those with close relationships to the political establishment. As discussed above, politics and patronage played an important role. Moreover, local organizations in some colonias enforced strict rules on permissible fuels, no doubt motivated in part by a desire not to be undercut by neighbors using cheap fuel. Other organizations, the CDP among them, actively opposed the push for propane use.

Our survey results suggest that formal and informal regulation played an important role in brickmakers' adoption decisions, but it is unclear which was more influential. Over a quarter of brickmakers who adopted propane cited "outside pressure" as the key reason. Over three quarters were aware of government regulations regarding fuel choice and, of these, the majority has seen the regulations enforced. Over half the brickmakers reported that local organizations influenced their fuel choices, and a third said that neighbors, as distinct from local organizations, did.

Education

FEMAP sought to educate brickmakers about the harmful effects of burning dirty fuels via one-on-one discussions with individual brickmakers and in organized training sessions. In addition, an effort was made to reach a broader audience by distributing a comic book on the health advantages of burning clean fuels, a project funded jointly by El Paso Natural Gas and the Southwest Center for Environmental Research and Policy. (Unfortunately, by the time this comic book was ready for distribution in January, 1995, the conversion effort already had been derailed by a rise in propane prices).

Judging from our survey results, these efforts have not had a great impact on brickmakers' perceptions of the private health benefits of propane use. Only about one in ten brickmakers in our sample associated any adverse health effects with brickmaking and, even more surprisingly, only about one in ten believed that firing with propane was "healthier" than firing with debris. Twenty-nine percent of the brickmakers surveyed believed that dangerous fumes were emitted when kilns are fired with propane. Rumors that propane is harmful were spread by brickmakers who opposed the adoption of propane. Nevertheless, these rumors may have some basis in fact, since, when propane lines and tanks leak and burners are not adjusted properly, noxious fumes can be emitted. Recently two brickmakers who used propane died. While it is far from clear that propane actually had anything to do with these deaths—the official causes were cirrhosis and a heart attack—persistent rumors to this effect reflect a prejudice that is not uncommon among the brickmakers.

Market Intervention

Project leaders pursued two strategies to reduce competitive pressures to cut costs by burning dirty fuels. First, in March, 1993, they helped to negotiate an agreement among leaders of all the major brickmakers unions (including the ones that opposed the propane program) to fix the price of bricks at 250 pesos per thousand.[42] The price floor was meant to be high enough to allow all brickmakers to afford propane. Predictably, however, many of the brickmakers who were still burning debris began to cheat, selling at prices below the agreed-upon floor, and the agreement soon collapsed. Second, project organizers tried to mandate a market for "ecological bricks" fired with propane. In 1993, FEMAP and the city government were able to get construction companies to agree to buy ecological bricks instead of dirty bricks and brick substitutes such as cinderblock. As noted above, in August, 1994, the federal government ordered INFONAVIT to use only ecological bricks.[43] Unfortunately, neither INFONAVIT nor the construction companies consistently complied with their agreements to use only ecological bricks, and the arrangement floundered.

Technological Change

Project participants devoted a great deal of effort to attempting to lower the (variable) costs of using propane by improving kiln fuel efficiency. Engineers from El Paso Natural Gas, Los Alamos National Laboratories, and, to a lesser extent, Grupo Peñoles and the Gas Research Institute in Chicago all have been involved. Many of the early experimental kilns required radical departures from traditional kilns, e.g., (1) an electric heat source and conveyor belt, (2) multiple propane burners inserted into the sides of the kiln, and (3) a multichambered kiln.[44] More recent prototypes use a traditional kiln complete with arches and a sunken firebox as a starting point. Engineers also have worked to develop low-cost measures for improving fuel efficiency, by modifying burners and changing the fuel mixture, the manner in which bricks are stacked, the way that the kiln opening is covered, and the way the bricks are dried prior to firing. Fifty-four percent of the adopters in our sample modified their kilns when they began to use propane, but in two-thirds of the cases, the modifications consisted of rebuilding arches or strengthening walls to enable them better to withstand the more intense heat generated by propane; energy saving was not a consideration. Thus, although the project did succeed in introducing technologies such as improved burners, and although it did persuade many brickmakers to change age-old production methods (both significant achievements in themselves), as of this writing, it had not yet developed and diffused affordable and low-technology innovations that significantly reduce the variable costs of using propane by improving fuel efficiency. In fairness, it must be said that not even the most efficient experimental kilns tested at ECO-TEC would make propane attractive at current relative prices.

Peak and Decline of the Brickmakers' Project

The high-water mark of the Brickmakers' Project, as measured by the percent of brickmakers using propane, probably occurred in the fall of 1993. The exact number of brickmakers who were using propane at that time is unclear. According to some estimates, fully 70 percent of the brickmakers in Ciudad Juárez were using propane. A 1994 FEMAP report states that, as of April, 1993, 55 percent of brickmakers were using propane. In our sample, which probably is biased toward brickmakers who were more likely to have used propane, 62 percent claimed ever to have used propane.[45]

With propane prices rising, by early 1994, brickmakers began abandoning propane in droves. In the course of our interviews in Ciudad Juárez in July, 1995, we found only one brickmaker who still used propane. This price increase created pressures which caused key participants in the Brickmakers' Project to defect. The most important defection was by the municipal government, which stopped enforcing a ban on burning debris in late 1993, removing the principal "stick" in the program and leaving only the "carrots," such as subsidies and patronage. Brickmaker organizations increasingly dropped out, as they were undercut by competitors using dirty fuels. Finally, as noted above, construction companies and the federal workers housing agency, which had agreed to purchase ecological bricks, reneged.

Looking Ahead

Despite the failure of the propane initiative, FEMAP continues to work on promising solutions to the brick kiln pollution problem and the Brickmakers' Project yet may prove successful in converting the brickmakers to cleaner fuels and more efficient kilns. Today, FEMAP effectively has given up on trying to convince the Ciudad Juárez brickmakers to use propane and now is promoting natural gas, which burns as cleanly as propane but is far less expensive. The main obstacle to its use is that natural gas requires considerable permanent infrastructure. Whereas propane can be delivered and stored in portable tanks, natural gas must be piped in. Also, expensive decompressors are needed at pipeline junctions. There are no natural gas pipelines in any of the colonias where the brickmakers are located, although in some there are pipelines nearby. FEMAP estimates that it would cost $800,000 to $1,000,000 to install all the necessary infrastructure, and currently is helping to seek funding from a variety of sources. Having conducted tests in several colonias, it plans to use federal funds remaining in the trust fund it administers to subsidize the adaptation of traditional kilns to natural gas. FEMAP also has been seeking to help brickmakers diversify into higher value products, such as "Saltillo tile" and roof tile, and to break into the U.S. market for these goods. FEMAP hopes that a higher selling price will enable the brickmakers to afford clean fuels.[46]

Engineers at El Paso Natural Gas, Los Alamos National Laboratory, and the University of Utah continue to push ahead with efforts to design more fuel-efficient kilns. At the University of Texas at El Paso, a project is under way to design solar kilns that would, in effect, partially cook bricks before they are fired, thereby reducing the time that the kiln needs to be fired.[47]

As for the brickmakers themselves, the failure of the propane initiative has left some cynical and disaffected. This is not surprising, given that some brickmakers incurred significant costs (in time as well as money) in switching to propane and then back to debris. Predictably, the failure of the initiative has provided confirmation for the local organizations which have opposed it all along.

LESSONS FOR CROSSBORDER ENVIRONMENTAL MANAGEMENT

The principal lessons of the Brickmakers' Project for crossborder environmental management entail: the promise of grassroots private sector–led initiatives, effective strategies for pollution control in the informal sector, and the fragility of voluntary market-based environmental initiatives in the informal sector.

The Promise of Grassroots Private Sector–Led Initiatives

In many respects, the Brickmakers' Project is a success story. It has attracted a remarkable amount of publicity and support. Although the diffusion of propane among the brickmakers was limited and temporary, it nevertheless represents a significant achievement in view of the obstacles involved, especially the reduction in propane subsidies (without this, propane use probably would have continued to grow). Thus the project illustrates that private sector initiatives hold considerable promise as means of addressing crossborder environmental problems, and informal sector pollution problems in particular. Moreover, the Brickmakers' Project illustrates that such initiatives enjoy a number of advantages over state programs. First, the project's binational composition suggests that private sector–led initiatives effectively can sidestep the bureaucratic and sovereignty disputes that often confound government attempts at international cooperation. Second, the enthusiasm that the project generated among funders, participants, and the public at large suggests that private sector–led initiatives may be able to draw more freely on public sympathy for border environmentalism than top-down bureaucratic initiatives.

But does the qualified success of the Brickmakers' Project imply that border environmental problems might best be left to private sector grassroots organizers? Definitely not, since, in all likelihood, the Brickmakers' Project would not have had as much success without unusually strong U.S. and Mexican federal support, the support of the municipal and state governments, and the leadership of a well-established, politically savvy nongovernmental organization.

As discussed above, U.S. and Mexican federal support for the border environment largely grew out of the NAFTA fight, coupled with the emergence of environmentalism as a Mexican electoral issue. Federal support took the form of funding, publicity, and pressure applied to state and municipal governments. FEMAP might not have been able to initiate or sustain a high level of effort without federal support. The same might be said of other participants in the project, including the University of Texas at El Paso, Monterrey Tech, El Paso Natural Gas, and Los Alamos National Laboratory. Federal support aside, it is doubtful that the project could have made as much progress in diffusing propane without the willingness of the Ciudad Juárez municipal government to crack down on the burning of debris.

Finally, FEMAP is not a typical grassroots organization. It enjoys strong political and business ties that have helped it attract federal support, convince other institutions to participate, elicit the cooperation of local governments, and generate publicity. A less well-established and well-connected organization would have had much more difficulty organizing such an effort.

Thus the first lesson of the Brickmakers' Project is that private sector grassroots binational initiatives can work—indeed, they may be more effective than public sector initiatives—but they require strong public sector support and some ability on the part of project organizers to leverage this support.

Effective Strategies for Pollution Control in the Informal Sector

As noted above, environmental management in the informal sector is inherently difficult: the number, size, dispersion, and anonymity of informal firms make them exceedingly difficult to monitor; intense competition biases them toward the lowest cost inputs; and poverty in the informal sector weakens political will to impose stiff compliance costs. Despite these obstacles, the Brickmakers' Project was, for a time, successful in inducing brickmakers in Ciudad Juárez to adopt a clean technology that in most cases raised production costs. What organizational strategies were responsible?

First, project organizers encouraged a cooperative relationship with the brickmakers, not the adversarial relationship that exists between most regulators and polluters. Instead of focusing solely on punishing nonadopters, project organizers sought to reward adopters by providing equipment, credit, technical extension, subsidies, and less visible patronage. Just as important, they worked to develop good relationships with brickmakers, both individually and as represented by various organizations. This cooperative approach in effect was built into the project from the beginning. FEMAP is a social service organization (dedicated to "improving the quality of life for individuals") and, as such, envisioned the project as a means of improving the lives of the brickmakers as well as the environment.[48] Indeed, some of the project's activities, such as management training and the recent effort to diversify the brickmakers' products, have been oriented primarily toward economic development, not environmental management. Also, as a nongovernmental organization with no enforcement powers, FEMAP had little choice but to adopt a cooperative approach.

FEMAP's cooperative approach helped to defuse opposition to stiff enforcement measures. FEMAP encouraged both formal and informal enforcement. It supported the municipal government's crackdown on nonadopters. The municipal government strengthened enforcement by setting up a peer monitoring system wherein city authorities responded to citizen complaints. Project organizers also encouraged informal enforcement by labor unions and neighborhood associations. This may have been an increasingly easy task, as adopters had an incentive to insure that their neighboring competitors, too, switched to propane.[49] One must note, however, that the success of the Brickmakers' Project in promoting both formal and informal enforcement depended largely upon the fact that neighbors could observe violations because they could see or smell toxic smoke. Other types of informal sector pollution, such as the dumping of waste oil into sewers by mechanics, would not be detected so easily.

While providing inducements and promoting enforcement seem to have succeeded in convincing brickmakers to adopt propane, efforts to introduce energy-efficient kilns, educate the brickmakers regarding private health and safety issues, and manipulate the market for bricks were less successful, because they were either poorly implemented or simply ill-conceived. The project's inability to design and diffuse innovations that significantly improve kiln fuel-efficiency seems to be due mainly to the difficulty of the task. In part, though, it reflects a failure to embrace two well-established principles for introducing new technologies in low-income settings. First, to the extent possible, intended adopters should participate in designing and building the innovation. Second, new technologies must be "appropriate"—that is, both affordable and consistent with existing levels of technology.[50] By contrast, most of the early experimental kilns were designed by highly trained engineers and involved radical departures from existing kilns.

Our survey results suggest that the Brickmakers' Project may have missed an opportunity to promote the adoption of propane by educating brickmakers about the private health benefits of doing so. The importance of proper use of propane also seems not to have been communicated effectively. As mentioned above, few respondents perceived burning propane to be "healthier" than burning debris.

It is clear that all of the project organizers' attempts to manipulate the market for bricks—by fixing the price of bricks in March, 1993, and, later that same year, attempting to mandate the use of ecological bricks—failed utterly. In most cases, contravening market forces in developing countries simply does not work; monitoring is too difficult, and cheating is too easy, especially in the informal sector.[51]

The Brickmakers' Project, then, suggests four lessons for environmental management in the informal sector. First, effective environmental management requires establishing a cooperative instead of an adversarial relationship with firms, one based on recognizing the socioeconomic needs of informal sector employees. Beyond rhetoric, establishing such a relationship translates into encouraging the participation of local unions and political organizations and, even more concretely, providing a variety of inducements to offset the costs involved in producing more cleanly, including subsidies on new inputs, credit, and technical extension. Second, environmental regulations can be enforced in the informal sector by relying on peer monitoring and informal regulation, as well as formal regulation. Third, new "green" technologies must be appropriate—that is, both affordable and low-technology. Finally, attempts to manipulate informal markets simply do not work.

The Fragility of Voluntary Market-Based Environmental Initiatives in the Informal Sector

Ultimately, FEMAP's propane program was undermined by steady reductions in propane subsidies on the U.S.-Mexican border. Does the history of the propane initiative's demise hold any lessons?

On the one hand, this history might be seen as evidence of a failure on the part of the Mexican government to coordinate conflicting policy initiatives. While the federal government actively supported and funded the effort to convert brickmakers to propane, it simultaneously supported the liberalization program that undermined it. But this liberalization program was part of a broad economic reform. The economic benefits of this reform well may have outweighed the costs, including the environmental costs. To reduce these environmental costs, the Mexican government might have subsidized propane use by key low-income users who were bound to substitute into dirty fuels. But such a policy would have been difficult to implement and likely would have perpetuated the black market in subsidized propane.

It seems equally unfair to fault the organizers of the Brickmakers' Project. Propane prices began to increase only in 1992. By this time, an initial group of the brickmakers in Ciudad Juárez already had switched to propane, and the project was completely organized around the strategy of engineering a switch. Today project organizers are promoting an alternative strategy—the adoption of natural gas.

Thus the overarching lesson to be learned from the demise of the propane initiative is somewhat bracing: in volatile developing economies, voluntary market-based environmental initiatives among informal sector firms (whose profit margins are exceedingly small) are bound to be fragile, even when well designed and implemented.

NOTES

1. The authors gratefully acknowledge the financial support of the Tinker Foundation in New York. For their cooperation, we also thank the officers of FEMAP, Octavio Chávez, Nancy Lowery, Francisco Núñez, Carlos Rincón, the TNRCC, Hubert Eldridge of El Paso Natural Gas Company, and the brickmakers of Ciudad Juárez. The opinions expressed in this paper are those of the authors alone and do not necessarily reflect the views of any of the organizations or individuals named above.
2. G. Ranis and F. Stewart, "V-Goods and the Role of the Urban Informal Sector in Development," Yale University, Economic Growth Center Discussion Paper No. 724 (1994), 18–20.
3. For the section on "El Paso–Ciudad Juárez Air Quality," our main sources were: C. Richard Bath and V. Rodríquez, "Comparative Binational Air Pollution Policy in El Paso, Texas, and Ciudad Juárez, Chihuahua," *Borderlands* 6, no. 2 (1983): 171–97; and Francisco Núñez, D. Vickers, and Peter Emerson, "Solving Air Pollution Problems in Paso del Norte," paper prepared for "Conference on the Border Environment," El Paso, Tex., Oct. 3 and 4, 1994, unpublished paper (1994). For the sections "Traditional Brick Making in Ciudad Juárez," "Strategies," and "Peak and Decline of the Brickmakers' Project," we relied mainly on our own survey research, described in detail in Allen Blackman and Geoffrey J. Bannister, "Community Pressure and Clean Technology: An Economic Analysis of the Adoption of Propane by Traditional Mexican Brickmakers," *Journal of Environmental Economics and Management* 35, no. 1 (forthcoming). For the sections entitled "Birth of the Brickmakers' Project," "Mexican Federal Support," "Participation in the Brickmakers' Project," and "Looking Ahead," we relied mainly on conversations with project participants and on FEMAP reports. For the section on "NAFTA and Mexican Politics," our principal sources were: Stephen Mumme, "Clearing the Air: Environmental Reform in Mexico," *Environment* 33, no. 10 (1991): 7–30; and N. Kublicki, "The Greening of Free Trade: NAFTA, Mexican Environmental Law, and Debt Exchanges for Mexican Environment Infrastructure Development," *Columbia Journal of Environmental Law* 19 (1994): 59.
4. Núñez, Vickers, and Emerson, "Solving Air Pollution Problems," 1.
5. U.S. Bureau of the Census, *County and City Data Book* (Washington,

D.C.: GPO, 1994); Mexico, Instituto Nacional de Estadística, Geografía e Informática, *XI Censo Nacional de Población y Vivienda* (Aguascalientes, Mexico: Instituto Nacional de Estadística, 1992).

6. T. Barry, *The Challenge of Crossborder Environmentalism* (Albuquerque, N.Mex.: Resource Center Press, 1994), 38–39; Bath and Rodríquez, "Comparative Binational Air Pollution Policy"; Conversation, Allen Blackman with M. Parra, air investigator, TNRCC, in El Paso, Tex., Mar. 21, 1996.
7. For history and analysis of efforts to control air pollution along the border, see Bath and Rodríquez, "Comparative Binational Air Pollution Policy"; and N. Johnstone, "International Trade, Transfrontier Pollution, and Environmental Cooperation: A Case Study of the Mexican-American Border Region," *Natural Resources Journal* 35, no. 1 (1995): 39–40.
8. Sawdust, scrap wood, pecan shells, and other combustibles generally are distributed to brickmakers by independent agents who collect the material from the maquiladoras. Often the maquiladoras pay these agents to take it away, enabling the agents to sell it at a low price.
9. E.g., A. Johnson, J. Soto, Jr., and J. Ward, "Successful Modernization of an Ancient Industry: The Brickmakers of Ciudad Juárez, Mexico," paper presented at the New Mexico Conference on the Environment, Apr. 25, 1994, 2; and M. Mendoza, "LANL [Los Alamos National Laboratory] Helping Mexico Clean Up Border Smog," *Albuquerque (N.Mex.) Journal*, Nov. 5, 1995. Though widely held, this hypothesis is undocumented. According to TNRCC, no emissions inventory ever has been performed for Ciudad Juárez.
10. Johnson, Soto, and Ward, "Successful Modernization," 7.
11. City of Ciudad Juárez, Dirección Municipal de Ecología, "Complaints According to their Origin," Jan., 1995, unpublished document, available from the authors or FEMAP.
12. FEMAP, "Summary of Brick Kilns Census," May, 1991, unpublished document, available from the authors or FEMAP.
13. Johnson, Soto, and Ward, "Successful Modernization," 2, 5.
14. The famous sunken park in Mexico City (El Parque Hundido) on Avenida Insurgentes—a large park that is two meters below the level of the street—was the result of just such excavations performed by a group of brickmakers who were moved out of the city 30 years ago.
15. FEMAP, "Summary of Brick Kilns Census."
16. These figures come from ibid. and from Blackman and Bannister survey. It is difficult to calculate the profit from informal activities precisely, owing to poor record keeping and the use of family labor.
17. Minimum wage figures are for March, 1996, at 18.7 pesos per day; "De regios: Leche da fuerte glope contra salario mínimo," *El Norte,* Chihuahua, Chihuahua, Mar. 19, 1996. The monthly figure is calculated assuming 26 days of work per month. Data are from Banco de México, *The Mexican Economy* (México, D.F.: Banco de México, 1996), 289, table 27.
18. Blackman and Bannister survey; FEMAP, "Summary of Brick Kilns Census."
19. Mumme, "Clearing the Air," 10; Kublicki, "Greening of Free Trade," 93.
20. According to Carlos Rincón, the El Paso Citizens' Environmental Advi-

sory Committee helped to popularize the statistic that traditional brick kilns are the third or fourth leading source of air pollution in El Paso–Ciudad Juárez. Conversation, Allen Blackman with Carlos Rincón, project director, EDF, in El Paso, Tex., Dec. 5, 1995.

21. Conversation, Blackman with Rincón, Dec. 5, 1995. Conversation, Geoffrey Bannister with E. Suárez, executive director, FEMAP, Dec. 11, 1995.
22. FEMAP, *Report for Dec. 1992–Jan. 1994*, unpublished (1994), 21.
23. U.S. Congress, Office of Technology Assessment, "The Border: A Boundary, Not a Barrier," in *U.S.-Mexico Trade: Pulling Together or Pulling Apart* (Washington D.C.: GPO, 1992), 123.
24. Mumme, "Clearing the Air," 27.
25. Kublicki, "Greening of Free Trade," 84.
26. FEMAP, *Report for Dec. 1992–Jan. 1994*, 15.
27. PRONASOL (Programa Nacional de Solidaridad) was a program initially administered within the office of the Presidency that offered matching funds to poor urban communities for the installation of basic infrastructure such as sewers and electricity. In April, 1992, PRONASOL was merged with the environmental agency (SEDUE) to create SEDESOL, a cabinet-level ministry.
28. FEMAP, "Why is This News?" ECO-TEC brochure, (undated), available from authors or FEMAP; Conversation, Bannister with Suárez, Dec. 11, 1995; FEMAP, *Report for Dec. 1992–Jan. 1994*, 8.
29. C. Irigoyen, "Environmental Plan Announced," *Diario de Juárez*, Feb. 13, 1992; FEMAP, *Report for Dec. 1992–Jan. 1994*, 14; "Commitment to Environment a Moral Issue, Not Political: CSG," *Diario de Juárez*, Aug. 6, 1993; FEMAP, "ECO-TEC Report for October, 1994–March, 1995" (1995), 4.
30. Conversation, Bannister with Suárez, Dec. 11, 1995.
31. "INFONAVIT Ordered to Use Environmentally Safe Brick," *Diario de Juárez*, Aug. 5, 1993.
32. It is interesting to note that, since Nov., 1992, both the governor of Chihuahua and the mayor of Ciudad Juárez have been members of PAN. Francisco Barrio Terrasas, the governor since 1992, was mayor of Ciudad Juárez in the early 1980s.
33. FEMAP promotional brochure (undated), available from the authors or FEMAP.
34. For a description of EPA-funded activities, see EPA, "Compendium of EPA Binational and Domestic U.S.-Mexico Activities," EPA 160-B-95-001, June, 1995.
35. See, e.g., J. Marcus, "The Brickmakers' Story," *Pipeliner*, published by El Paso National Gas Co., El Paso, Feb., 1994.
36. Propane prices were provided by FEMAP, and the estimates of the per-brick energy costs are based on Blackman and Bannister survey data.
37. FEMAP, "ECO-TEC Report for December, 1993–August, 1994" (1994), 3, 4; Conversation, Bannister with Suárez, Dec. 11, 1995. ECO-TEC has tried to recover costs—a requirement for all FEMAP projects—by selling the bricks manufactured in its experimental kilns to construction companies and by selling its extension expertise to Mexican states and municipalities.

38. M. Cruz, "Brick Producers Plan Formation of Cooperatives," *Norte de Ciudad Juárez,* Feb. 8, 1993.
39. Federal authorities technically assumed responsibility for point sources of emissions within 100 kilometers of the U.S. border, under the terms of the La Paz Agreement. In practice, however, federal participation in day-to-day enforcement efforts—at least with regard to brick kilns in Ciudad Juárez—was limited until 1995. EPA, *Environmental Protection Along the U.S.-Mexican Border,* EPA 160-K-94-001 (Washington, D.C.: EPA, 1994), 3.
40. Conversation, Blackman with Rincón, Dec. 5, 1995.
41. FEMAP, *Report for Dec. 1992–Jan. 1994,* 12.
42. Ibid.
43. Conversation, Bannister with Suárez, Dec. 11, 1995.
44. FEMAP, "ECO-TEC Report for December, 1993–August, 1994," (1994); Los Alamos National Laboratories, "Mexican Brick Kiln Study, Mar. 18–20, 1994," Report LAUR-94-1322, (1994). Johnson, Soto, and Ward, "Successful Modernization."
45. Our sample probably is biased toward propane users, because we interviewed brickmakers who happened to be working at their kilns when we administered our survey. These were most likely to be active, relatively wealthy, large-scale brickmakers—the same type most likely to adopt propane. Also, we did not interview brickmakers in two distant and isolated colonias, Anapra and Fronteriza Baja, where, by all accounts, few if any brickmakers ever used propane.
46. Conversation, Bannister and Blackman with F. J. Alfaro Mata, director of ECO-TEC, in Ciudad Juárez, Apr. 24, 1995; Conversation, Allen Blackman with Carlos Rincón, July 21, 1995.
47. For a description, see EPA, *Environmental Protection Along the U.S.-Mexican Border.*
48. FEMAP, brochure, (undated), 1. The two objectives of the Brickmakers' Project were: (1) "To help reduce environmental pollution . . . [and] (2) To help improve standards of well-being and quality of life for brickmaking families and others like them, by preserving production capacity, modernizing their small enterprises, improving productivity and product quality, and increasing income through avoidance of intermediaries"; FEMAP, *Report for Dec. 1992–Jan. 1994,* 2.
49. In this sense, competition among the brickmakers seems to have worked in favor of the project. This suggests that, if enough firms can be brought on board by hook or crook, eventually competition will ensure that the adoption of a cost-increasing clean technology becomes self-perpetuating.
50. See, e.g., D. Barnes et al., "The Design and Diffusion of Improved Cooking Stoves," *World Bank Research Observer* 8, no. 2 (1993): 119–41; F. Stewart, *Technology and Underdevelopment,* 2d ed. (London: Macmillan, 1977), 1–3.
51. See, e.g., D. Lal, *The Poverty of Development Economics* (Cambridge, Mass.: Harvard Univ. Press, 1985).

7

CARBÓN I/II

AN UNRESOLVED BINATIONAL CHALLENGE

MARY KELLY

Some days, a visit to Big Bend National Park, the largest protected area along the United States–Mexico border, isn't what it used to be. The park is located in the Chihuahuan desert at the bend of the Rio Grande/Rio Bravo in the southwestern corner of Texas. It long has been famous for spectacular views, as well as high mountain ranges, rich biological diversity, and whitewater rafting opportunities. On clear days, from the top of the Chisos Mountain range, visitors can see up to 150 miles into Mexico, across the beautiful Maderas del Carmen mountain range in northern Coahuila.[1] These vistas and regional air quality are now, however, increasingly threatened by uncontrolled sulphur dioxide emissions from two large power plants in northeastern Mexico and by emissions from several other sources in Texas and northern Mexico.

In the last few years, these magnificent views, which help attract over 330,000 annual visitors to the park,[2] increasingly have been obscured by a thin white haze. Visibility measurements taken by the National Park Service in the summer of 1995 show, for example, that, during the period of August 21–27, visibility averaged less than twenty-three miles.[3] Average summer visibility at the park usually is about fifty miles. On August 22, visibility was nine miles, the lowest ever recorded that could be attributed to air pollution.

As is discussed in more detail in this chapter, most scientists attribute the summer visibility problems at Big Bend National Park largely to sources of air pollution in northeastern Mexico, particularly heavy industries in Monterrey and Nuevo León, and one very large source of sulphur dioxide, the Carbón I and Carbón II coal-fired electric generating plants. These are located near Piedras Negras, Coahuila, about 125 miles to the southeast of the park. Air pollution sources in Texas also may be having a detrimental effect on visibility in the Big Bend area, especially during fall and winter months.[4]

But it is not only Big Bend National Park that is adversely affected by this air pollution. In the early 1990s, the State of Texas completed acquisition of the 276,000-acre Big Bend Ranch State Natural Area, just upriver from the park. In addition, a 196-mile stretch of the Rio Grande below the park is designated a Wild and Scenic River under U.S. federal law—the only river in Texas with such a designation. The 101,500-acre Black Gap National Wildlife Refuge is located on the eastern edge of the park; and, in January, 1996, the Texas Parks and Wildlife Department received, through the efforts of the Texas office of the private Conservation Fund, a donation of 40,000 acres in the Chinati Mountains, located to the west of the park.[5]

Even more significant, on the day before he left office, Mexico's former President Carlos Salinas de Gotari declared the establishment of two new protected areas across from Big Bend: Maderas del Carmen and Cañon de Santa Elena, with a combined size of 1.2 million acres.[6] Management plans, which are necessary to make these protected areas a reality, recently have been prepared.[7] These decrees were a critical first step in fulfilling long-standing hopes of a binational set of protected areas to maintain and showcase the region's incredible beauty and biodiversity.[8]

Together, all these protected areas, along with substantial existing private conservation efforts on both sides of the border, could provide an important basis—maybe the best basis—for sustainable long-term development in this arid part of the border region, development based upon the conservation of the natural environment. The U.S. National Park Service estimates that, in 1994, Big Bend National Park alone contributed over $36.7 million to the income of local area businesses or individuals, from sale of goods and services to park visitors; generated almost $2.5 million in tax revenues; and accounted for over twelve hundred local jobs.[9] These positive economic impacts could be expanded greatly and extended to the Mexican side of the border with the establishment of the Maderas del Carmen and Cañon de Santa Elena reserves.

Increasing regional air pollution, however, threatens to undermine this promise. This chapter explores the failure of the U.S. and Mexican governments to remedy one of the largest and most obvious individual sources of regional air pollution affecting the Big Bend–Maderas del Carmen area—the Carbón I and Carbón II power plants. The Carbón I/II situation illustrates many aspects of the classic transboundary pollution controversies explored in this book, including:

- The role of federal and state governments.
- The role of private investors.
- The role of environmental nongovernmental organizations (ENGOs).
- Disparity in environmental standards and financial resources to control pollution.
- Conflicts between environmental protection and job creation or retention.
- The relative importance of NAFTA, the NAFTA Environmental Side Agreement, and the Commission for Environmental Cooperation (CEC).
- The absence of a clear framework through which the U.S. and Mexico might better resolve transboundary pollution disputes.

ORIGINS OF THE CONTROVERSY

The Carbón I and Carbón II power plants are fired primarily with locally mined low-grade coal, supplemented by some coal imported from the U.S.[10] The power plants are owned and operated by the Comisión Federal de Electricidad (CFE), an agency of the Mexican federal government. With a combined capacity of 2600 megawatts at full capacity, the plants provide relatively low-cost electricity to municipal and industrial customers in northeastern Mexico, particularly in the states of Nuevo Leon and Coahuila.

The 1200-megawatt Carbón I began operating at near full capacity in the early 1980s. The plant is not equipped with either particulate matter or sulphur dioxide control equipment.[11] When Carbón I was being planned and constructed, apparently no crossborder consultation took place between the U.S. and Mexican governments about the potential effects of air pollution from the plant. Nor does there seem to have been any involvement by ENGOs.[12]

Carbón II was planned along with Carbón I, and the generating units were purchased early on. Actual construction on the Carbón II plant, however, did not begin until the late 1980s. Carbón II is equipped with electrostatic precipitators (ESPs) for control of particulate matter but does not have any control for sulphur dioxide emissions.

Because of their size and lack of emission control equipment, the Carbón I/II plants are a significant source of air pollution. Information from state and federal government agencies indicates that, at full capacity, the plants would produce about 250,000 tons of sulphur dioxide per year, making them the largest source of air pollution within five hundred kilometers of Big Bend National Park.[13] According to a Texas Natural Resource Conservation Commission (TNRCC) analysis of the plants, the emissions from Carbón I/II will cause violations of Class I visibility standards for Big Bend National Park. If the plants were built in the U.S., they would be the nation's seventh largest source of sulphur dioxide emissions.[14] Sulphur dioxide emissions, which consist in part of small sulfate particles, are major contributors to the type of haze which has decreased visibility in the Big Bend region.[15] Sulphur dioxide and sulfate pollution also can damage sensitive vegetation.[16]

Based on air quality modeling and other studies, the National Park Service concluded that the impact of Carbón I/II exceeds the combined impacts of the twenty-eight largest sources of sulphur dioxide emissions in Texas, even though these Texas sources together emit slightly more sulphur dioxide than the two Carbón plants.[17]

Air quality and visibility reduction in national parks and the relationship of air quality to pollutants emitted from coal-fired power plants are, of course, not new issues. For example, millions of dollars and countless hours have been spent in analyzing and trying to resolve visibility degradation at the Grand Canyon National Park in Arizona. This degradation often is attributed in part to the operation of coal-fired power plants in the "Four Corners" region of New Mexico.[18]

New electric power plants in the U.S. are required to meet strict air pollution control standards.[19] Mexico's standards for such plants are substantially less stringent. For example, depending on density and heating value of the coal being burned, Mexico's regulations limit sulphur dioxide emissions to between 4.4 and 8.76 pounds per million BTUs,[20] while U.S. standards limit sulphur dioxide emissions to 1.2 pounds per million BTUs.[21] Mexico's regulations for power plants limit particulate matter emissions to 0.31 pound per million BTUs, but the U.S. regulations are about ten times as stringent, limiting particulate emissions to 0.03 pound per million BTUs for new coal-fired power plants.[22]

In June, 1993, the *Washington Post* ran an article on the Carbón II plant and its possible relevance in the debate over the environmental impacts of NAFTA.[23] The trade agreement then was a heated topic of debate in the U.S. Congress. The article discussed the potential environmental impact of the plants, based on information obtained from the EPA and the National Park Service. These agencies were just beginning to look into the potential effects of the power plants on Big Bend National Park and air quality in the Texas-Mexico border region. The article also discussed ongoing negotiations between CFE and private investors for privatization of the Carbón I/II plants. A consortium of investors, including the Grupo Acero del Norte (GAN), a powerful industrial group from Coahuila, and Mission Energy, a subsidiary of Southern California Edison, a California utility, had been negotiating an agreement to privatize ownership and operation of the plants. There also was some discussion of partial financing for the privatization deal being provided by the International Finance Corporation (IFC), the private financing wing of the World Bank.

In the months following the *Post* article, ENGOs began to take a much closer look at the Carbón I/II situation.[24] They urged the EPA to work with Mexico to get an agreement on pollution control for at least the Carbón II plant. Most groups believed that the best alternative—given the dependence of local jobs on the Micare mine supplying coal to the plants—would be for the Mission Energy and GAN consortium to retrofit scrubbers to the plant to control sulphur dioxide emissions, thereby reducing both localized health impacts and the regional visibility degradation affecting the Big Bend–Maderas del Carmen area. Concerted efforts also were made to highlight the role of Southern California Edison, Mission Energy's parent corporation, which, while being one of the utility industry's more progressive companies in respect to environmental issues in the U.S., was willing to own and operate a plant without sulphur dioxide control just thirty miles south of the U.S.-Mexico border.

In October, 1994, however, shortly before the final votes on NAFTA, the Mission Energy and GAN partnership withdrew from negotiations with CFE.[25] While a full analysis of the withdrawal never was made public, it appears that the negotiations were affected not only by the environmental controversy, but also by opposition to privatization in some quarters at CFE.[26]

CAUGHT UP IN THE NAFTA DEBATE

As is discussed in more detail elsewhere in this book, the environment, and in particular the environment along the U.S.-Mexico border, were central issues in the NAFTA debate. In this vein, Carbón II became a fairly high-profile issue.[27] For example, in September, 1993, the House Committee on Energy and Commerce, Subcommittee on Energy and Power, held a hearing on the energy-related aspects of NAFTA, focusing on the Carbón II issue. Although concerned with the potential effects of the plant, the U.S. administration found itself in a defensive posture: it was trying to insure Congress that NAFTA would not lead to more situations like Carbón I/II and that the agreement would, in fact, foster cooperation in resolving such issues. At the same time, discussions between EPA and its Mexican counterpart, SEDESOL, had not produced any concrete results.

In the end, however, the endorsement of the NAFTA Environmental Side Agreement by several large U.S. environmental organizations somewhat neutralized the environmental issues, including Carbón II, in the final debate.[28] Ironically, as is noted in more detail below, the NAFTA Environmental Side Agreement does not provide clear remedies for a problem like Carbón II. The side agreement does allow citizens of the three NAFTA partner nations to file complaints about a government's failure effectively to enforce its own environmental laws.[29] The Carbón II plant, however, complies with Mexico's air pollution regulations; the problem is weak standards, something not addressed directly by the NAFTA side agreement.[30]

POST-NAFTA: PROMISES OF INCREASED COOPERATION FAIL TO MATERIALIZE

After NAFTA was approved, then Gov. Ann Richards of Texas, a NAFTA proponent and a longtime fan of Big Bend's rafting and other recreational opportunities, tried mightily to leverage the promise of increased U.S.-Mexico environmental cooperation into action on Carbón II. Backed by analyses from TNRCC, the governor attracted public attention to the issue[31] and wrote to high-level Mexican government officials expressing her concerns. For example, in a June 30, 1994, letter to Herminio Blanco, head of SECOFI, Governor Richards pointed out, "Unless this matter is resolved . . . you and I will have failed to demonstrate that the NAFTA or other U.S.-Mexico agreements provide effective mechanisms for addressing binational environmental problems."

With some limited success, the State of Texas tried to insert itself into discussions between EPA and SEDESOL (now SEMARNAP) about the Carbón II issue. According to Diana Borja, director of the Office of Border Affairs at TNRCC, the state still feels that this issue is very much under federal control.[32]

The governor's overtures, which were accompanied by entreaties from former Texas Lt. Gov. Bill Hobby,[33] and continuing interest on the part of the Texas General Land Office in alternative fuel options for Carbón II,[34] nevertheless were soundly rejected by Mexico.

Two other developments during this time are noteworthy. First, parallel to the Carbón I/II controversy, a fight had been brewing over a proposed coal strip mine near Eagle Pass, Texas, which is just across the border from Piedras Negras. The proposed mine was to be owned and operated by a company called Dos Repúblicas. It was opposed vigorously by several adjacent landowners and the Lone Star Chapter of the Sierra Club. The landowners were concerned about potential adverse impacts of the proposed mine on their property and groundwater resources. The Sierra Club supported the landowners and raised two additional objections: the effect of the mine on habitat for the endangered ocelot and jaguarundi cats and the proposed end use of the coal to be mined—firing the Carbón II power plant.

The proposed mine needed permits from EPA's Region VI, the TNRCC, and the Railroad Commission of Texas, which regulates strip mines. The landowners and the Sierra Club unsuccessfully fought the state permits.[35] The Sierra Club and other ENGOs also participated in EPA's environmental impact statement (EIS) process for the proposed mine, raising the Carbón II end-use issue. Despite its own well-documented and public position that the coal-fired units of Carbón I/II were related to visibility degradation at Big Bend National Park, EPA rejected arguments for denying the Dos Repúblicas federal permit.

In its final EIS, EPA stated that, while it agreed that impacts from Carbón I/II on Big Bend air quality indeed were severe, it would not deny the discharge permit, because "impacts from Carbón I/II will occur regardless of coal source, [thus] permit denial will not solve the problem, and EPA does not have the authority to prohibit export of U.S. resources which cause the country environmental harm . . . EPA believes that the U.S. policy should be to take actions which will generate the investment capital needed to directly solve the Carbón I/II problem."[36]

A second and parallel development was EPA's move to take the Carbón II controversy out of the public view by relegating it to a "technical working group" of staff from EPA and their counterparts in Mexico's federal environmental agency.[37] This working group took almost a year to agree that the prevailing summer wind direction was from southeast to northwest, which would move Carbón I/II air pollution toward the Big Bend–Maderas del Carmen area.[38] At the same time, EPA tried to convince Mexico to use an experimental sulphur dioxide–removal technology called the ADVACATE system on Carbón II and sweetened the offer with $500,000 in financial assistance.[39]

Perhaps understandably, Mexico rejected these overtures. In a May 18, 1995, letter to U.S. Ambassador to Mexico James Jones, SEMARNAP Secretary Julia Carabias and Energy Secretary Ignacio Pichardo stated that Mexico would prefer to wait for further tests on the ADVACATE system, then being installed on a Polish generating plant, before considering its use on the Carbón II facility. In the same letter, Mexico rejected any conclusion that Carbón I/II were adversely affecting air quality in the Big Bend region, claiming that the available data were inadequate to support such a conclusion. The two ministers did express a willingness further to study "the impacts of sources in both countries" and to engage in discussions about "the present and future implications of industrial development in both countries."[40]

In the fall of 1995, in preparation for President Ernesto Zedillo's state visit to the U.S., federal officials north of the border made some efforts to place the Carbón II issue on the discussion agenda and to include some sign of progress in the final statement resulting from the meeting. Despite these efforts, the Carbón II issue was not addressed in the public statement released after Zedillo's October visit.[41]

In late March, 1996, a meeting of the binational coordinators under the La Paz Agreement produced a joint U.S.-Mexico statement on the Carbón II issue.[42] This statement appears to acknowledge the possible impact of Carbón I/II emissions on air quality in the Big Bend–Maderas del Carmen area but provides only for further regional air quality studies. At this writing, results from the first phase of air sampling, which was completed in the fall of 1996, have yet to be released, even though analysis of the samples was completed in the spring of 1997. Funding for further studies—including tracer studies on the emissions from Carbón I/II—is uncertain.

Thus, over four years after NAFTA's approval and three years after the Carbón II issue hit the radar screen, there has been virtually no progress in resolving the conflict, and air quality in the Big Bend–Maderas del Carmen region continues to deteriorate.[43]

INVOLVEMENT OF THE CEC

At first blush, the Carbón I/II controversy might seem ideally suited for resolution by the trinational Commission for Environmental Cooperation (CEC) created by the NAFTA Environmental Side Agreement. It is a transboundary pollution problem of significant dimensions, the solution of which requires close cooperation at the highest federal levels. Moreover, there could be an important trade link, if Mexico were to begin exporting electricity from the Carbón I/II complex across the border through the Texas grid.[44] Other potential North American trade aspects include crossborder export of electricity, and import of U.S. or Canadian natural gas as a potential substitute fuel.

Crossborder Energy Transmission Issues

The crossborder energy transmission issues associated with Carbón I/II or other electric generating facilities in the northern border area of Mexico are complex. At this point, there is very little export of electricity from Mexico to the U.S. and no major transmission lines for such export exist. Under Annex 602.3(5)(c) of NAFTA, however, both U.S. and Mexican governments must permit CFE to negotiate deals allowing independent power producers in Mexico to sell electricity to users in the U.S.[45] This could apply to the Carbón I/II plant, if it ever were fully privatized. Before NAFTA was finalized, some analysts theorized that independent power producers might come under "scrutiny under NAFTA, leading to decisions on fuel supply based on air emissions, with [natural] gas presumably favored."[46]

Moreover, the EPA has stated that the new power lines required for import of electricity into the U.S. likely would require an environmental assessment under the National Environmental Policy Act,[47] and that such an assessment would need to examine related air quality impacts associated with generation of the electricity to be exported.[48] Nevertheless, because neither NAFTA nor its Environmental Side Agreement directly addresses the issue of production process standards,[49] it is unclear whether the U.S. could prohibit import of electricity from a plant that was causing air quality degradation in the U.S.[50] Such an effort would be governed by disciplines established by the General Agreement on Tariffs and Trade (GATT) and the World Trade Organization, which still are subject to multiple interpretations.[51]

These aspects of the issue, in fact, have been referred to by those who sought to bring the Carbón I/II controversy to the CEC. For example, Jan Gilbreath, a researcher at the Lyndon B. Johnson School of Public Affairs, University of Texas, Austin, urged the council of the CEC to look into Carbón I/II, in her public comments at the council's first public meeting in July, 1994.[52] Others, including the Environmental Defense Fund and the Texas Attorney General's Office, also made inquiries about a possible CEC role.

In response to these inquiries, the CEC secretariat set up an informal working group of experts from the three NAFTA countries to explore the Carbón II issue from a trinational perspective of long-range pollutant transport. This working group never made progress on the Carbón I/II issue, however.[53]

Reluctance concerning CEC involvement in the Carbón I/II issue has been expressed in some quarters, including by the author of this chapter. Concern has centered on whether a specific binational border issue is appropriate for resolution by a trinational body such as the CEC, especially given the lack of a defined framework for resolving such issues.

Nevertheless, a number of aspects of the NAFTA Environmental Side Agreement potentially are relevant to the Carbón I/II situation, including Article 13 (secretariat reports), Articles 14 and 15 (citizen submissions), and the side agreement's preamble and objectives.

Under Article 13, in addition to the required annual report, the CEC secretariat is empowered to "prepare a report on any other matter within the scope of the annual program."[54] The secretariat also may prepare a report on any "other environmental matter related to the cooperative function of the Agreement." For this latter option, however, the secretariat must give thirty days notice to the council, which can veto the secretariat's request by a two-thirds vote.

It was this Article 13 authority upon which the secretariat relied for its investigation of, and report on, a massive migratory bird die-off in the Silva Reservoir in Guanajuato, Mexico. This report was well received in most quarters, and the secretariat received high praise for the objectivity and timeliness of the report.[55]

There are limits to the Article 13 process, however. First and probably foremost, there is no obligation on the part of the governments to implement any recommendation that might be contained in an Article 13 report. If an Article 13 report really were directed at a serious reduction in the pollution coming from the Carbón I/II plants, the most obvious recommendation would be the use of flue gas desulphurization technology, but already there is a firm stalemate between the countries on this avenue. In this instance, it comes down to political will—the will of the U.S. government to find funding to retrofit the scrubbers, and the will of the Mexican government to acknowledge that these plants threaten air quality on both sides of the border and should be brought up to a more reasonable standard of pollution control.

Second, with respect to the Carbón I/II controversy, unless the secretariat could demonstrate that the matter was within the scope of the annual work program,[56] the council, by a two-thirds vote, could prohibit the secretariat from moving forward. Given the complexity of the issues and the apparent political sensitivity of the Carbón I/II matter, there would be a very good chance of two votes against an investigation by the secretariat.

Third, the use of Article 13 for the Silva Reservoir case was ideal for two reasons: (1) the cause of the die-off was uncertain, but, with proper investigation, it was readily discernible; and (2) specific, generally viable recommendations for avoiding future problems could be (and were) made by the secretariat. In the Carbón I/II situation, however, much of the modeling and analysis already has been done (even though there are some disagreements over the results). The solution, while clear, right now is politically unpalatable, no matter where the recommendation originates.

Article 14 allows the secretariat to "consider a submission from any non-governmental organization or person asserting that a party [NAFTA country] is failing to effectively enforce its environmental law."[57] As noted earlier, however, the Carbón II plant complies with Mexico's air quality regulations, and so it is difficult to conclude that there is a readily apparent basis for an Article 14 claim. Moreover, there is very little basis to argue that EPA is failing effectively to enforce the U.S. Clean Air Act with respect to Carbón II by not taking some more definitive action, since there is no clear mandate or authority for EPA to do so.[58]

Finally, there is language in the preamble to the Environmental Side Agreement and in the agreement's Article 1 (Objectives) that could be a basis for some level of CEC involvement. The preamble "reaffirms" the three countries' commitments to the principle that activities in one country should not harm the environment of the other.[59] And, arguably, all of the objectives stated in Article 1[60] favor speedy resolution of a controversy such as that posed by the Carbón II situation.[60] The problem, of course, is that the NAFTA Environmental Side Agreement, like most international legal mechanisms, is very short on mandatory mechanisms to achieve its objectives. Like the 1983 La Paz Agreement, the NAFTA side agreement still is heavily reliant upon an informal cooperative framework—a framework which can work well for general issues but one which leaves much to be desired in the resolution of specific difficult issues.

LESSONS OF THE CARBÓN I/II DILEMMA

The Carbón I/II controversy provides a variety of insights into, and lessons for, U.S.-Mexico cooperation on transboundary pollution problems. These insights and lessons are important for several reasons. First, despite the sidestepping by both governments, the Carbón I/II complex is a clearly identifiable major source of transboundary pollution. From a technical perspective, the problem is fairly well defined and straightforward,[61] albeit difficult to resolve. In some ways, then, the Carbón I/II situation, if it were to be resolved successfully, could provide an extremely important precedent for future disputes.

Second, some sources have indicated that the Mexican government apparently is planning Carbón III. If the controversy is not resolved soon, there will be no guidance for necessary pollution controls on Carbón III.

Third and maybe most important, resolution of the Carbón I/II issue is necessary to insure an important basis for future sustainable development in the arid, binational Chihuahuan desert region of the Texas-Mexico border. By committing to protect air quality and visibility in the Big Bend–Maderas del Carmen–Cañon de Santa Elena protected area region, the governments would be committing themselves to protect an economic development asset of immeasurable value—one that could be the key to a sustainable future for the region. If air quality is preserved, the protected areas will attract hundreds of thousands of visitors annually, providing employment and economic opportunity for many residents and establishing a framework for compatible economic development activities throughout the region.

The Carbón I/II situation starkly illustrates the weaknesses of our current U.S.-Mexico binational cooperative framework. In most areas, the 1983 La Paz Agreement, the mainstay of the binational environmental relationship, provides only a broad and vague mandate for informal cooperation and discussion. Except where specific annexes have been negotiated through lengthy diplomatic discussions, the La Paz Agreement does not contain any measures which require either the U.S. or Mexico to take specific steps to prevent activities which may result in adverse environmental or public health impacts in the other country.[62]

Even more clearly, Carbón I/II illustrates that NAFTA, by itself, is not necessarily going to increase U.S.-Mexico cooperation on other issues, especially where there are substantial differences of opinion or priorities between the two countries. Disparity in environmental standards and available financial resources can be, as they have been in this situation, difficult issues to tackle in binational discussions that proceed without clear obligations, under an informal framework.

Some inherent, and maybe desirable, limitations of the CEC also are brought to light by Carbón I/II. The situation helps to draw clear distinctions between binational border issues that must be resolved by the two federal governments, acting in concert with appropriate state and local governments, on the one hand, and issues that are more appropriate for trinational decision making or investigation or that have a direct NAFTA trade link, on the other hand.

For governments, the Carbón I/II situation could be interpreted to show that, by taking the issue out of public view—i.e., by discussing it only in closed binational "working groups"—much of the motivation to overcome the inertia of doing nothing but study the issue is lost. Public awareness and public pressure are critical in any situation where there are substantial financial or policy implications involved in resolving a transboundary pollution dispute.

Finally, the Carbón I/II case holds at least two key lessons for ENGOs. The first is the need to recognize the full binational dimensions of the situation. That is, often there are environmental and economic development implications for both sides of the border in transboundary pollution situations. These implications need to be linked more closely in the public actions and statements of ENGOs who are pressing for resolution of the problem.

The second lesson is that all parties—including ENGOs—have to be more willing to talk about where the money will come from to address the problem. While the most appropriate solution to the problem is likely to be retrofitting scrubbers to at least Carbón II to reduce sulphur dioxide emissions, the public debate too often has ignored the issue of how pollution controls can be financed. The most viable option, of course, would be for the U.S. to put together a combination of loans and grants for the necessary equipment costs, thus cushioning the financial impact on Mexican ratepayers. This would be no easy task, but it never will come about if the organizations which want the problem resolved are not willing to push hard for the financing.

Given the new government focus on air quality studies, instead of finding funds for scrubbers for Carbón I/II, it may be time for ENGOs north of the border, in particular, to take a much closer look at the Texas sources impacting Big Bend–Maderas visibility during winter months. Another important factor will be ENGOs' ability to maintain and reinforce attention to the trade-environment linkage in the Carbón I/II situation.

NOTES

1. National Park Service (NPS), "Pollution Plagues Big Bend National Park," news release, Sept. 1, 1995, available in author's files or at Big Bend National Park. Information on visibility and air pollution problems at Big Bend, including photos, is available through the Internet at http://www.aqd.nps.gov/ard/.
2. NPS, *State of the Park Report: Big Bend National Park, Rio Grande Wild and Scenic River* (Big Bend National Park: NPS, 1996).
3. NPS, "Pollution Plagues Big Bend."
4. NPS conducted a regional haze analysis for Big Bend National Park, examining the impacts of 28 major Texas sources of sulphur dioxide within 480 kilometers of Big Bend National Park. Based on these modeling results, the NPS found that "emissions from the Texas sources likely cause significant impairment to visibility at the park. It is estimated, based on 1988 meteorological data, that these existing sources cause perceptible visibility impairment between 20 and 34 times during the year depending on preexisting conditions (average or good visibility) at the park. The majority of these impacts occur during the winter and fall seasons, which are the seasons with the best visibility and highest visitation at the park. Based on model predictions, visibility at the park can be reduced by as much as 40 to 50 percent." John P. Christiano, chief, Air Quality Division, NPS, to Jodena Henneke, director, Air Quality Planning Division, TNRCC, Nov. 30, 1994. The NPS did note that this modeling was based on allowable, not actual, emissions of sulphur dioxide from the Texas sources, but emphasized that if Big Bend visibility was to be protected (as required by the Clean Air Act), the allowable emission limits for the Texas sources must be reduced. Ibid.

In addition to the NPS research, several university researchers, as well as TNRCC, have begun investigations into the various sources of air pollution affecting this area.

5. "State Agency Given 40,000 Acres in West Texas," *Victoria (Tex.) Advocate*, Jan. 26, 1996.
6. "Maderas Decree," *Diario Oficial de México*, Nov. 7, 1994, 11–20. The decrees establishing these new protected areas base the designation on the need to protect endangered species (e.g., black bear and peregrine falcon), sensitive ecosystems, biological corridors, the Rio Grande/Rio Bravo watershed, and important archeological sites. In addition to abundant wildlife and over seventy species of nesting birds, the Maderas del Carmen area has many features likely to attract visitors, including spectacular vistas, high mountain forests, and perennial springs. In the Cañon de Santa Elena area, attractions include rugged desert hiking, wildlife viewing, and the pine and oak forests of the Sierra Rica.
7. Richard Lowerre, "The Serengeti of the Texas-Mexico Border?" *Ambiente Fronterizo* (Austin: Texas Center for Policy Studies), Mar., 1996, discusses new protected areas in Mexico and current environmental threats affecting their viability.
8. Conservationists and government officials in both countries have discussed the concept of a binational set of protected areas in this region since Big Bend National Park was first established in 1944. Laurence Parent, "Will Mexico's Sierra del Carmens Become a Companion to Big Bend National Park?" *Texas Parks and Wildlife* 48, no. 4 (Apr., 1990): 4–11. For an interesting discussion of some challenges facing a binational set of protected areas on the U.S.-Canada border, see Joseph L. Sax and Robert B. Keiter, "Glacier National Park and Its Neighbors: A Study of Federal Interagency Relations," *Ecology Law Quarterly* 14 (1987): 207, 237–40. The article discusses binational efforts to protect Glacier National Park from a proposed coal mine north of the border, including the role of the IJC.
9. NPS, *State of the Park*, 2.
10. "Carbón II Units Scheduled to Come on Line," *Latin America Power Watch*, Dec. 14, 1994.
11. Chris Green and Mary Kelly, *The Carbón II Dilemma: A Case Study of the Failings of U.S.-Mexico Environmental Management in the Border Region* (Austin: Texas Center for Policy Studies, Sept., 1993).
12. The La Paz Agreement, which requires consultation between U.S. and Mexican governments on major transboundary sources of pollution, was not in place until 1983. *Agreement Between the USA and the United Mexican States on Cooperation for Protection and Improvement of the Environment in the Border Area*, 22 ILM 1025 (1983). In 1987, the U.S. and Mexico signed Annex IV to the La Paz Agreement, in which the two countries agreed to specific reductions and controls on sulphur dioxide emissions from copper smelters in the Arizona-Sonora border region. This is one of the most successful crossborder cooperative efforts regarding air pollution. Such ENGOs as the Border Ecology Project, based in Bisbee,

Ariz., were an important catalyst in pushing for this annex. Similarly, in the El Paso–Ciudad Juárez region, the EDF has been a leading force behind the recent creation of the International Air Quality Management District and crossborder cooperation to reduce this airshed's severe pollution problems. Kevin G. Hall, "U.S., Mexican Officials Likely to Form Air Quality Pact," *Journal of Commerce,* Mar. 27, 1996; "U.S. and Mexico Combine Forces to Fight Pollution," *Wall Street Journal,* Mar. 29, 1996.

13. NPS, *Big Bend National Park Carbón I and II Power Plants* (Fact Sheet), Feb., 1995. Emission estimates for the Carbón I/II complex vary, in part because data on the actual sulphur content and other properties of the coal being burned in the plants has not been readily available.

14. TNRCC, Memo to Gov. Ann Richards regarding Carbón I/II power plants, June 30, 1994, author's files. "Class I" visibility standards are set by the Clean Air Act and are designed to prevent degradation of visibility in pristine areas such as national parks. See Clean Air Act, 42 USC §§ 7470, 7491.

15. National Research Council, *Protecting Visibility in National Parks and Wilderness Areas* (Washington, D.C.: National Research Council, 1993). NPS, *Visibility Protection,* available on the Internet at http://www.aqd.nps.gov/.

16. World Bank, *World Bank Environmental Guidelines* (Washington, D.C.: World Bank, 1984), 427–28.

17. Dora A. Tovar and Jan Gilbreath, *Clean Energy Along the Texas-Mexico Border: Building Consensus in Protecting Air Quality* (Austin: U.S.-Mexican Policy Studies Program, LBJ School of Public Affairs, Univ. of Texas at Austin, Aug., 1995).

18. NPS, *Protecting Visibility*. Information on Grand Canyon visibility issues can be obtained through the Grand Canyon Visibility Transport Commission, which is staffed through the Western Governors' Association (tel. 970-623-9378).

19. Green and Kelly, *Carbón II Dilemma,* 4–5.

20. The British Thermal Unit (BTU) is a measure of the heating value of fuel.

21. TNRCC to Gov. Richards, June 30, 1994. On several aspects of the Carbón II issue, see Michael Shapiro, acting asst. administrator for air and radiation, EPA, to Cardiss Collins, chair, Subcommittee on Commerce, Consumer Protection, and Competitiveness, U.S. House Committee on Energy and Commerce, Aug. 7, 1993, author's files.

22. Shapiro to Collins, Aug. 7, 1993. It should be noted that these figures are based on Mexico's standards as they applied to the Carbón I/II plants. Mexico is undertaking a process to revise many of its environmental standards.

23. Tod Robberson, "Cloud Over Trade Pact—Texas Too," *Washington Post,* June 22, 1993.

24. See, e.g., Green and Kelly, *Carbón II Dilemma*. ENGOs besides Texas Center for Policy Studies actively involved in the Carbón II issue at that time included the Sierra Club, Public Citizen, Greenpeace USA, and EDF.

More recently, the National Parks and Conservation Association has joined the fray and taken several steps to urge resolution of the Carbón II controversy.

25. Andy Pasztor, "SCE Corp. Drops Mexican Electric Plant under Political, Environmental Pressure," *Wall Street Journal,* Oct. 12, 1993.
26. Ibid.
27. See, e.g., Andy Pasztor, "Power Plants in Mexico Cast Pall Over NAFTA," *Wall Street Journal,* Sept. 8, 1993; Ralph K. M. Haurwitz, "NAFTA Environmental Concerns," *Austin (Tex.) American-Statesman,* Oct. 4, 1993.
28. See Steve Charnovitz, "The NAFTA: Green Law or Green Spin?" *Law and Policy in International Business* 26 (Fall, 1994): 1–77.
29. North American Agreement on Environmental Cooperation among USA., Canada, and Mexico, 32 ILM 1480 (1993), Arts. 14 and 15.
30. The preamble to the NAFTA environmental side agreement does reaffirm the responsibilities of the three countries to "ensure that activities within their jurisdiction or control do not cause damage to the environment of other states or of areas beyond the limits of national jurisdiction." North American Agreement on Environmental Cooperation (NAAEC), "Preamble" (Washington, D.C.: EPA, 1993).
31. See, e.g., "Power Plant Concerns Richards," *Austin American-Statesman,* Apr. 9, 1994; Mike Leggett, "Richards Gets Back to Nature at Big Bend," *Austin American-Statesman,* Apr. 10, 1994; Robert Bryce, "Big Bend Acid Rain," *Texas Observer,* May 6, 1994, 9–10.
32. Interview, Mary Kelly with Diana Borja, Director of Border Affairs, Austin, Texas, Apr. 2, 1996.
33. Bill Hobby, "NAFTA Offers Hope for Clean Border," *Austin American-Statesman,* May 16, 1994. Governor Hobby expressed concern about potential damage from emissions to an expensive new observatory being constructed at McDonald Observatory in Jeff Davis County, just north of Big Bend National Park.
34. Texas Land Commissioner Garry Mauro was one of the first state officials interested in the Carbón II issue, from the perspective both of air quality protection and of marketing natural gas to México for use in the Carbón II plant. See elsewhere in this chapter for more on the natural gas option.
35. At the Texas Railroad Commission, a hearings examiner did recommend, based upon evidence presented during a contested hearing on the stripmine application, that the permit be denied. That recommendation was overturned by the full three-member commission. TNRCC has issued the wastewater discharge permits required for the Dos Repúblicas operation, although the Sierra Club has appealed that decision in U.S. district court.
36. EPA, Region VI, *Final Environmental Impact Statement: Eagle Pass Mine, Maverick County, Texas,* EPA No. 906/01-95-001, Jan., 1995, C-51.
37. Remarks of James Yarborough at the Conference on Clean Energy Along the Texas-Mexico Border, LBJ School of Public Affairs, Univ. of Texas at Austin, Feb. 17, 1995 (author's notes).
38. Ibid. By contrast, the NPS has taken many appropriate actions to docu-

ment visibility degradation at the park and to educate both the general public and park visitors about the problem. NPS, *State of the Park,* 3.

39. Tovar and Gilbreath, *Clean Energy.* The ADVACATE system, developed by researchers at the Univ. of Texas, Austin, relies on a lime-based duct injection technology to reduce sulphur dioxide emissions by up to 70 percent, at a reported one-third to one-half the cost of conventional wet-scrubbing technology. The World Bank has financed a demonstration of the technology on a 120-megawatt power plant in Poland, but the ADVACATE system is not yet in use in any full-scale power plant.

40. Julia Carabias, Secretaría de Medio Ambiente, Recursos Naturales y Pesca, and Ignacio Pichardo, Secretaría de Energía, to U.S. Ambassador to Mexico James Jones, May 18, 1995, author's files.

41. President Zedillo stated that he and President Clinton had "spoken about our border, and we agreed to make it clean and safe and to make it an opportunity for productive activities and well-being." Transcript of remarks by President Clinton and President Zedillo of Mexico, state visit, Washington, D.C., Oct. 10, 1995, 40.

42. As of this writing, the statement has not yet been made available to the public.

43. This conclusion was reaffirmed in the author's interviews of various key personnel in the U.S. government in 1996, each of whom preferred not to be quoted.

44. Green and Kelly, *Carbón II Dilemma,* 6.

45. NAFTA, Dec. 17, 1992, 32 ILM 289, ch. 6, Annex 602.3(5)(c).

46. Michelle M. Foss, Francisco G. Hernández, and William Johnson, "The Economics of Natural Gas in Mexico Revisited," *International Association for Energy Economics Journal* 14 (1993): 17, 40. This article provides an excellent discussion of issues related to natural gas exploration, development, and use in Mexico, including export and import aspects, in a post-NAFTA context. While conversion of Carbón II to natural gas fuel was an option explored by Mission Energy, the Texas General Land Office, and Texas natural gas producers, the significantly higher price of natural gas (relative to locally mined coal) and NAFTA's very favorable tariff treatment for coal imported into Mexico (ibid., 39–40) make such a conversion for Carbón II economically unattractive, esp. at the levels of natural gas firing required to achieve significant reduction in sulphur dioxide emissions.

47. 42 USC § 4321ff.

48. Shapiro to Collins, Aug. 7, 1993, p. 5.

49. Production process standards are standards or regulations that restrict the import of products based upon how the product was produced, rather than product characteristics. The most widely discussed U.S. standard of this type is the restriction on importation of tuna caught in a manner unsafe to dolphins.

50. See Charnovitz, "NAFTA: Green Law or Green Spin?" 35–38, for a good discussion of production process standards.

51. Ibid.
52. Jan Gilbreath, Remarks to the Council of the North American Commission on Environmental Cooperation, Washington, D.C., July 26, 1994.
53. The CEC did recently issue a report on long-range air pollutant transport in North America, though that report did not deal specifically with the Carbón I/II issues. See *Continental Pollutant Pathways; An Agenda to Address Long-Range Transport of Air Pollution in North America* (Montréal: CEC, 1997).
54. NAFTA Environmental Side Agreement, art. 13.
55. A full description of the Silva Reservoir case is beyond the scope of this chapter. The CEC Secretariat's report is available on the Internet at http://www.cec.org.
56. The 1996 Work Program for the CEC does contain a project on "North American Air Monitoring and Modeling," under which at least a generalized analysis of the Carbón I/II controversy might fall. CEC, Annual Program and Budget 1996, Montréal, Nov. 21, 1996.
57. NAFTA Environmental Side Agreement, art. 14. See also art. 15 (preparation of a Factual Record by the Secretariat in response to a citizen submission); art. 45(1) (definition of failure to "effectively enforce"); and art. 45(2) (definition of "environmental law"). See also CEC, *Guidelines for Citizen Submissions,* available on the Internet at http://www.cec.org, also CEC, *Guidelines for Submission on Environmental Matters Under Articles 14 and 15 of the NAAEC* (Montréal, undated). In 1996 the CEC Secretariat accepted its first citizen submission—a complaint from Mexican ENGOs that Mexico had failed to effectively enforce its environmental law in issuing a permit for a new dock near unique coastal reefs off the island of Cozumel. Howard LaFranchi, "New Role for NAFTA: Saving Fish?," *Christian Science Monitor,* Jan. 23, 1996. The Secretariat earlier had rejected two submissions from binational coalitions of ENGOs who complained of the U.S.'s failure effectively to enforce environmental laws through the use of various congressional budget riders. See Jay Tutchton, "The Citizen Petition Process Under NAFTA's Environmental Side Agreement: It's Easy to Use, but Does It Work?" *Environmental Law Reporter (News and Analysis)* [Washington, D.C.: Environmental Law Institute] 26 (Jan. 1996): 10031.

    For more information on citizen submission procedures, see Karl Spiecker, Andrea Durbin, and Harry Browne, *A Citizens' Guide to NAFTA's Environmental Commission* (Washington, D.C.: Friends of the Earth and Albuquerque, N.M.: Interhemispheric Resource Center, Feb., 1995) and Texas Center for Policy Studies, *NAFTA's Environmental Side Agreement: A Review and Analysis* (Austin: Texas Center for Policy Studies, Sept., 1993).
58. Green and Kelly, *Carbón II Dilemma,* 12–13. This report contains a more detailed discussion of provisions of the federal Clean Air Act applicable to Carbón I/II. There may exist an Article 14 claim that neither EPA nor the TNRCC is effectively enforcing the Clean Air Act with respect to protect-

ing Big Bend National Park's air quality from degradation due to emissions from Texas sources. See n. 4 above.

59. This statement reflects a well-established principle of international law arising out of, among other things, the infamous Trail Smelter case, a 1941 international arbitration between the U.S. and Canada over sulphur dioxide emissions from a zinc smelter in Canada. *Trail Smelter Case* (U.S. v. Canada), 3 R. Int'l Arb. Awards 1950 (1941). See also "Transboundary Pollution from Mexico: Is Judicial Relief Provided by International Principles of Tort Law?" *Houston Journal of International Law* 10 (1987): 105.

60. The objectives of the environmental side agreement to NAFTA are to (1) foster protection and improvement of the environment in all three countries; (2) promote sustainable development; (c) increase cooperation among the countries better to "conserve, protect and enhance the environment, including wild flora and fauna"; (4) support the environmental goals and objectives of NAFTA; (5) avoid creating trade distortions or new trade barriers; (6) strengthen cooperation on the improvement of environmental laws and regulations; (7) enhance compliance with, and enforcement of, environmental laws; (8) promote transparency and public participation in the development of environmental laws and policies; (9) promote economically efficient and effective environmental measures; and (10) promote pollution prevention policies and practices.

61. This is not to say that resolution of the problem will not require large amounts of money, with attendant implications for both countries, or that Carbón I and II are the only sources causing visibility degradation in the Big Bend–Maderas del Carmen area. Clearly, there are other significant sources of air pollution, in both Texas and northeastern Mexico, that must be controlled better for full, long-term air quality protection.

62. For a detailed early analysis of the La Paz Agreement, see Mark A. Sinclair, "The Environmental Cooperation Agreement Between Mexico and the U.S.: A Response to the Pollution Problems of the Borderlands," *Cornell International Law Journal* 19 (1986): 87. On application of the La Paz Agreement in the Carbón II case specifically, see Green and Kelly, *Carbón II Dilemma*.

8

THE MAQUILADORAS AND THE ENVIRONMENT

ANN C. PIZZORUSSO

Much of the publicity surrounding the Mexican maquiladora (assembly plant) industry focuses on its environmental irresponsibility. Tales of toxic material disappearing into the desert, hazardous waste being dumped into streams, and contaminants polluting soil and water are ubiquitous and, according to many, unhappily true. Mexico's environmental image has been disparaged in *Time,* the *New York Times,* and the television networks. The media have cited illegal dumping of toxic materials by unscrupulous businesses, while praising local citizens who are tracking down these polluters. There also has been massive media coverage on anencephaly, a mysterious disease occurring only in the border zone, which causes children to be born with part of their brain missing. However, stories of companies which do properly dispose of toxic material and do institute effective management systems rarely are covered. Unfortunately, this biased coverage produces a stereotype; all the maquiladoras are tarred with the same broad brush. In fact, the situation is mixed. And, while difficult problems exist, the trend is toward better environmental management for these border industries.

Part of the problem is that Mexico has promulgated stringent environmental regulations while unable to develop its environmental infrastructure fast enough to allow for compliance. Without facilities in place to handle hazardous materials, toxic wastes are being disposed of improperly on remote parcels of land or, more commonly, dumped into streams and rivers. This activity is extremely lucrative for dishonest contractors, because, in the absence of a fully developed and effective enforcement regime, many businesses still are unwilling to pay to have these materials disposed of properly. The disposal contractors have minimal costs if they dump the waste illegally, unbeknownst to the innocent business owner who hired the contractor.

This situation adds to the pollution problem and gives all business a bad name.

With budgetary constraints hampering Mexican enforcement and the United States' assistance limited by sovereignty, what can be done to protect the environmental integrity of our neighbor? Since pollution does not recognize borders, what can be done to assure the health and safety of all persons living in the region? The answer is corporate responsibility. Each company benefiting from maquiladora incentives has an obligation to insure that it is environmentally responsible.

This is easier said than done. For American companies operating in the border zone, environmental responsibility is a top priority. Unfortunately, environmentally responsible companies are operating within the same geographical area as companies which are not so responsible, giving all companies a negative image while handing the polluters an apparent short-term advantage. Environmentally responsible companies operating in the border zone therefore find themselves in a no-win situation. They are operating at a higher standard, which costs more money, yet they are berated for alleged environmental disregard because of the actions of their neighbors. The situation is exacerbated by the fact that the border zone encompasses a large geographical area with multinational implications. This area's land, water, and air is being subjected to pollution which sometimes originates from the U.S. and moves into Mexico, and at other times the reverse is true. Urban areas, especially those with twin cities straddling the border, share not only pollution problems but common health concerns.

Mexico's desire to become more environmentally progressive resulted from its decision in the mid-1980s to open its economy and become a major player in world trade. Developing its economy and becoming a partner in the North American Free Trade Agreement (NAFTA) would require a more proactive position on the environment. There is no policy to create pollution havens. Mexico does not seek to develop on the basis of dirty industry, which would be a trap with respect to competitiveness and adopting new technologies. During 1989, Mexico, in cooperation with the U.S., promulgated more hazardous waste laws than in the past forty years combined. The laws designed for the maquiladoras were formulated to insure that participating companies, the hazardous waste materials they use, and the hazardous waste they generate are regulated jointly by the U.S. under its Environmental Protection Agency (EPA) and Mexico under its Secretariat of Environment, Natural Resources, and Fisheries (SEMARNAP), the successor to SEDESOL and SEDUE.[1]

All U.S. companies operating in the border zone are painfully aware of how costly cleanup of industrial sites can be. In the U.S., regulations have required the expenditure of hundreds of billions of dollars to correct problems resulting from past operations. Therefore, responsible corporations are instituting procedures, often on a worldwide basis, to minimize the risk of future environmental liability. It makes sense to institute responsible environmental practices in any manufacturing facility, regardless of the country. The fact that local enforcement may be less stringent—for whatever reasons—is no excuse to jeopardize the health and safety of workers or future profits, which sooner or later will have to be spent on remediation. This philosophy is common in most multinational corporations (especially U.S. companies, for in the U.S. liability for pollution falls upon the polluter).[2]

However, the efforts of these environmentally responsible companies have been slow to make any major impact on other businesses. Many companies, local and foreign, will seize an opportunity to cut costs by handling environmental obligations in the manner that costs the least. This often means operating illegally. Offers by multinational corporations to train personnel from small and mid-sized companies on environmental issues, recommend environmental hardware which would reduce pollution, or join with the local government to form cooperative efforts, largely have been ignored. The costs associated with these programs are too great for many small companies to bear, so the multinational firm ends up operating among companies with less environmental integrity. It remains to be seen if the current decidedly mixed situation on the border, home to over two thousand manufacturing plants (maquilas), will evolve toward an industrywide commitment to good environmental management. As discussed below, sooner or later this is likely to happen.

Responsible environmental care can make a difference, as has been the case, for example, in the maquiladoras of Philips Electronics North America Corporation. With fourteen manufacturing facilities and seven thousand employees in the Mexican border zone, Philips has a significant investment in the area.

Philips believes that a corporate environmental ethos must be formulated by top management and implemented by each employee. Philips's border zone environmental management program, which has evolved over the past eight years, demonstrates how a well-structured program, implemented by dedicated employees at all levels, can develop into one which can withstand the vicissitudes of economic and political flux. With its integrity and flexibility, the Philips border zone environmental program is a model which can be adapted anywhere in the world.

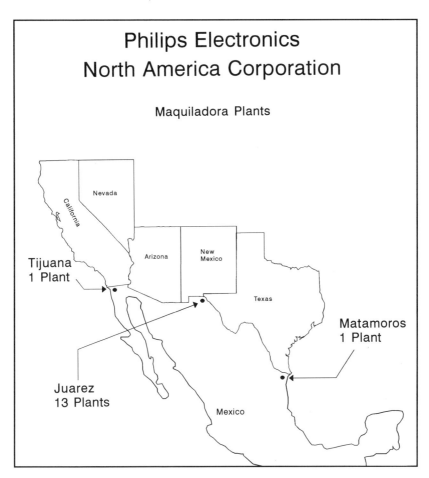

Maquiladora plants operated by Philips Electronics North America Corporation. Map by North American Philips Corp.

With the introduction of the ISO 14000 Standard, which is dedicated to the self-regulated measurement of environmental programs, corporations are under increasing pressure to adopt environmental management programs based on international standards, including Mexico.[3] To explain and promote ISO 14000 in Mexico, private sector workshops have been held in that country, starting in January, 1996, under the joint sponsorship of the U.S. Council for International Business (USCIB), the Canadian Council for International Business, and the Confederation of Mexican Chambers of Industry (CONCAMIN). In addition, the Montréal-based Commission for Environmental Cooperation (CEC) of NAFTA, with its mandate to enhance capacity building and cooperation, as well as to enforce environmental regulations, signed a memorandum of understanding supporting this training program. In fact, these private sector organizations and the trinational CEC regard the ISO 14000 system as an integral part of environmental management in North America.[4]

As discussed below, the border zone environmental management program of Philips is readily adaptable to ISO 14000 Standards. The firm's goal is to have most of its plants ISO 14000 certified by the year 2000.

Following the introduction of Mexico's comprehensive 1988 General Law for Ecological Equilibrium and Protection of the Environment, Philips initiated its own environmental management program in response to an increase in requests from Philips maquiladora plants for compliance guidance. At the time (1989), environmental staff at the company's New York headquarters had limited experience in Mexican environmental matters, so assisting the fourteen plants was a formidable task. We decided that the best immediate response was to organize a two-day environmental training course in El Paso for technical and operational personnel. Technical, regulatory, and legal issues applicable to maquiladora operations were addressed. We wanted plant managers to understand from the outset the importance of compliance, as they held the key to the personnel and monetary resources necessary to implement both the initial program and the recommended followup actions. No longer could managers foist these matters off on their environmental staff—the older way of doing things, which could lead to delays while first the managers and then all the employees bought into the new program.

The seminar could encompass all of the maquiladora plants because, unlike the U.S., where regulations vary from state to state, Mexico has a uniform permitting process. This allowed our training and deadlines for compliance to be standardized. At the seminar, an overview of the law and applicable regulations was presented. Each required form was reviewed, line by line, so there was an understanding of the paperwork and its purpose. Since the maquiladoras are affected by both U.S. and Mexican regulations, we debated the question of whether the plants should operate at the U.S. standard, where it exceeded the Mexican requirements. Management soon became convinced that compliance with the more stringent U.S. standard was sound environmental practice, would insure health and safety, and also would demonstrate our dedication to being responsible corporate citizens. Further, since it likely was only a matter of time before standards on both sides of the border would be similar, if not identical, future compliance costs could be reduced substantially if higher standards were adopted immediately.

Once the framework of the environmental program was developed, a permanent task force was formed. Plant managers agreed that, if Philips's headquarters in New York would commit resources and technical input, local management would participate fully and actively in using that expertise. Based on this understanding, the Border Zone Task Force (BZTF) was then established.

At the initial BZTF meeting, the first operational license application was completed by all plants. This set the pace for monthly meetings at which various permits, applications, and other documents were completed. Remarkably, within eight months, all Philips plants were in administrative compliance. The monthly meetings served as a forum for addressing, in addition to regulatory compliance, a myriad of related issues, from transportation to health and safety to recycling and waste minimization. Experts were invited to make presentations at these meetings concerning their particular areas of compliance. Having heard of our innovative approach, a representative from the EPA also attended a meeting. The meeting location was rotated monthly, so that each Philips maquiladora facility could be visited and evaluated by other members of the group. This allowed each member of the task force to become familiar with all operations and assess environmental compliance progress at each one.

Since 1989, the BZTF has evolved and accomplished much. This is because it was driven by a dedicated team of environmental professionals who did not stop at compliance but identified many areas where they could be proactive in the company and the community. Their attitude reflects the corporation's dedication to worldwide responsible environmental care. Since all environmental activities within the Philips organization are governed by the guiding principles of the Philips Electronics Policy Statement (reprinted at the end of this chapter), the Mexican operations, like their counterparts in the U.S. and elsewhere, strive to embody these principles.

As a result of its accomplishments, the BZTF came to the attention of SEDUE, then the federal environmental agency, and Philips staff members were invited to visit Arq. René Altamirano Pérez, a SEDUE official, in Mexico City to discuss their program. He was most impressed and used the program as a model of what conscientious corporate action in the maquiladoras could accomplish. Philips, in turn, offered to share its program with other maquiladoras.

Communication is an integral part of the Philips Mexican compliance program. Every month, each of Philips's Mexican environmental coordinators receives three newsletters. *Noticias de Philips* addresses issues of compliance with Mexican law, while *Environmental Infomemo* reports on U.S. environmental issues. Finally, *The Ad-*

visor, which is prepared by the occupational safety and health manager, disseminates information on health and safety issues. Thus environmental professionals not only can keep abreast of Mexican regulations, but also can follow U.S. trends and voluntary programs and implement them as well, while deepening the commitment of employees at all levels of the operation.

As of this writing, resource conservation is not as heavily regulated in Mexico as it is in the U.S. However, Philips's Mexican facilities always have placed a high priority on recycling, pollution prevention, waste reduction, and energy conservation programs. Recycling has been one of the BZTF's goals for many years. Innovative joint recycling ventures were formed by Philips facilities and Mexican companies. Two of these, formed by Philips Consumer Electronics and its suppliers, together with recyclers of styrofoam and solder, are models of small business development.

Two waste streams, styrofoam and solder dross, were causing disposal problems for our consumer electronics facilities in Ciudad Juárez. Both of these materials were recyclable but could not be treated in Mexican facilities due to the regulations requiring that wastes be treated in the country of origin of the raw materials. An innovative solution involving joint-venturing with a Mexican firm was developed by one of our employees at the television plant. At this facility, which manufactures over two thousand televisions per day, the large number of tubes brought in comes packaged in styrofoam, generating more than two tons of waste per day. Only two solutions were available: donate the styrofoam for building materials, or recycle it. Some of the styrofoam was donated under a company-sponsored program to assist self-help housing, but not all of it could be handled in that way. Recycling the styrofoam in the U.S. seemed the only way to dispose of it. As a last resort we contacted PetroPac, a Mexican-owned startup company which proposed to recycle this maquiladora waste for other uses. Our styrofoam provided them with a good baseline supply to begin their business. PetroPac then applied to SEDUE for a permit to recycle, and included Philips Consumer Electronics Company on their permit. The permit, and the inclusion of Philips as an integral part of the permit, was duly approved. Philips has been recycling styrofoam for several years now. We have saved money on disposal and are even receiving a small amount of money from the recycling operation. Most important, our scrap styrofoam is being recycled, and a Mexican-owned business was started.

With several wave solder machines, as well as single solder stations, Philips Consumer Electronics Company produces quite a bit of solder waste. Again, due to the regulations, this waste was classified by Mexico as hazardous waste and had to be transported to the U.S. for recycling. One of our employees began a search for a Mexican firm which could supply solder material as well as recycle the waste. A suitable firm, Soldaduras Omega, was contacted. Omega agreed to supply us with all of our solder at a much lower cost than our U.S. vendor, and to remove our solder waste (and recycle it into new products) at no cost. Working together not only allowed us to solve the challenge of a large amount of waste, but also enabled us to make a positive contribution to small business development.

Our Mexican operations also have adopted EPA's "Green Lights" program, a voluntary program to replace existing lighting systems in industrial facilities with energy-efficient technologies. According to EPA, Philips is the first corporation to take the Green Lights program to Mexico. EPA's 33/50 program, another voluntary effort, also has been implemented in Philips's Mexican facilities. The goal, to achieve 33-percent reduction of seventeen chemicals by 1992 and 50-percent reduction in the same chemicals by 1995, has been reached by Philips in all of its border zone plants.

Overall, the compliance program in Mexico is part of the corporation's worldwide environmental program. Philips Electronics set worldwide chlorofluorocarbon (CFC) reduction goals when the Montréal Protocol was signed, including a goal for North America. In our U.S. plants, elimination of CFCs has been achieved. In Mexico, however, Philips achieved CFC reduction even more quickly. In early 1990, Philips presented a report detailing its CFC reduction programs to Arq. Altamirano Pérez of SEDUE. The report noted that the border zone facilities of Philips had achieved a greater than 90-percent reduction in the use of CFCs, in one case pioneering a fluxless soldering process to do so. Altamirano used Philips's report in a presentation he made at a London meeting on the Montréal Protocol. By the end of 1990, the use of CFCs in Philips's maquiladoras was eliminated.

An outreach program to educate the Mexican public on environmental matters has been an element of Philips's Mexican environmental program from its inception. Philips's Consumer Electronics facility in Ciudad Juárez created a program to help provide shelter and water for residents of the colonias (substandard housing communities in the border zone, which are not served by standard utilities). Waste materials, such as wooden and metal pallets, cardboard, wood, and styrofoam, are used to build homes for those who otherwise would be living in cardboard shacks. Used water jugs are filled with potable water and distributed to colonia dwellers who otherwise would have limited access to such resources or would risk using contaminated containers. Workers from the facility donate their time to build houses and distribute water.

It bears repeating that much of what has been communicated to the public about environmental conditions and regulations in Mexico has come from the media; little good news or progress toward environmental compliance ever is reported. Philips representatives have been educating the public by speaking at seminars, meetings, and gatherings on topics ranging from basic Mexican environmental regulations to a discussion of Philips's Mexican environmental compliance program. In addition, Philips has been active in many trade and professional organizations, such as the USCIB, the Electronics Industries Association, the National Association of Manufacturers, the Association of Maquiladoras, and others. Six of the Ciudad Juárez facilities, including Componentes Eléctricos de Lámparas (Philips Lighting) and five of the facilities of Products de Consumo Electrónicos de Philips (Philips Consumer Electronics), were awarded SEDESOL's Green Flag Award in 1993. This award recognizes the achievement of total compliance for a three-year period. Only thirty-six awards were given to all facilities in Ciudad Juárez in 1993, of which Philips garnered almost one-fifth.

Philips also has been active in the multistakeholder Paso del Norte Air Quality Task Force, which designed and promoted an innovative plan locally to manage the air shared between Ciudad Juárez and El Paso. The chapter by Carlos Angulo and his colleagues discusses this issue at length. Suffice it here to say that participating in this successful effort at local institution building was consistent with Philips's proactive approach to environmental management. When, as expected, the plan is approved by officials in Mexico City and Washington, the private sector and Philips will be represented on the advisory committee to the new Paso del Norte Air Quality Management Basin.

During the 1993 debate surrounding the passage of NAFTA, environmental issues on the border became major causes of concern for the U.S. public. Many congressmen feared that lax enforcement in Mexico would lead to an acceleration of environmental abuses. Mexico insisted that it would enforce its environmental regulations and not become a dumping ground for industry. Philips supported NAFTA and continues to do so. To back up its endorsement, Philips could showcase an environmental management program that had been in place for many years, and which had achieved an impressive track record of compliance and proactive accomplishments. As a result of our work with industry organizations, such as the USCIB and the

Electronics Industries Association, we were able to show that many other U.S. companies were responsible corporate citizens where environmental matters were concerned. Based on these environmental achievements, a compelling argument for supporting NAFTA was made during the hard-fought battle for congressional approval in 1993.

Now, however, border environmental management programs confront new challenges and risks, with the January, 1995, peso devaluation and sudden near-collapse of Mexico's economy, continuing attacks on a now-embattled EPA, and cost cutting in the U.S. Congress. How have these events affected Philips's environmental management programs in Mexico? The good news is that the programs have remained intact, with the same high level of performance. This is due to a number of factors, the most important being that the program had an intrinsic integrity. That is, it has been supported throughout the organization; from the chairman to the line worker, each has regarded it as her or his responsibility to make it work. Next, compliance issues were dealt with early on, so that, over the years, the program changed from reactive to proactive. This took the form of voluntary programs and self-imposed goals adopted by the BZTF.

There are many benefits to this type of program: less risk, as issues are identified and dealt with; a better understanding of regulations because the process is ongoing; and greater efficiency as protocol is established. Finally, after the initial monetary outlay, there are economic benefits that accrue year after year, as increased environmental efficiency translates into a cost-effective method of operating. Eliminating hazardous materials, recycling, and retrofitting facilities for energy efficiency are all items that translate into dollars ultimately saved—all while protecting the environment. This result confirms a basic tenet of environmentalism: good practices produce good business.

As we look toward the year 2000, Philips has established a set of environmental goals for its worldwide operations. From Jakarta to Juárez, plants will be measured on how successfully they meet these standards:

- 25 percent reduction in energy consumption by the year 2000.
- 15 percent reduction in packaging by the year 2000.
- Implementation of certifiable environmental management systems, with ISO 14000 as a goal.
- Ecodesign as an integral part of the product development process.
- Environmental communication network both internally and externally.
- International life-cycle assessment.
- Supplier environmental partnering.

All products will be evaluated to determine their useful life, recyclability, and environmental status if they are shipped from one country to another. Supplier environmental partnering will continue to be an integral part of the program, in order to conserve resources and draw upon other points of view in determining how products can be manufactured in an environmentally sound manner. Philips's environmental partnering in the border zone has been quite comprehensive. In many ways it serves as a model for other countries.

It bears repeating that the environmental management system in Mexico is predicated upon achieving ISO 14000 certification. A training session for all our Mexican facilities was held early in 1996. It is gratifying to report that, due to the existing environmental management program, Philips's Mexican facilities are some of the front-runners in North America for rapid ISO 14000 certification. Indeed, our lamp manufacturing facility in Ciudad Juárez was ready to be ISO 14000 certified in the fourth quarter of 1996. Because there were no certified auditors ready to work in Mexico, the facility had to wait until 1997, causing it to be the second plant in Mexico to be ISO 14000 certified, rather than the first. Four of our other Mexican facilities followed suit and were certified in 1997: Philips Lighting in Monterrey and two consumer electronics plants and a lamp manufacturing plant in Ciudad Juárez. Interestingly, in the corporation, Mexico leads the way with four certifications, whereas only one Philips plant in the U.S. is certified.

The ISO 14000 certification process offers all businesses in Mexico the opportunity to adopt an environmental management program, because ISO 14000 is a system which starts off with the basics and can be expanded. This system is ideally suited for small to mid-sized businesses which can start their programs with modest investments and build upon them. Will ISO 14000 serve the current needs of Mexican manufacturers? Probably not, because the ISO 14000 is a program known more widely in multinational corporations, where the quality standard ISO 9000 is firmly in place. Once again, worldwide companies are adopting programs which have many benefits: competitiveness, environmental responsibility, and standardization worldwide. That these companies are adopting the standard places increasing pressure on all companies, regardless of size, to adopt the ISO standards. This change in focus, by which environmental considerations move from being primarily regulatory-driven to being an integral part of a corporate strategy, is important. Those companies which incorporate this broader vision will be at a competitive advantage while improving environmental conditions. Those who don't will be at a competitive disadvantage if customers insist on doing business with environmentally responsible companies.

Indeed, a recent article entitled "Maquilas Save Money, Environment," profiled two plants in Matamoros—Philips's Airpax de México and Summit Componente.[5] Both had opened their facilities to the Environmental Defense Fund (EDF) for an environmental assessment. (Recall that McDonald's stopped using styrofoam packaging after a similar evaluation by EDF.) Among EDF's recommendations were lighting changes, plastic waste recycling, and reusable packaging—all of which Philips is now implementing. Having independent organizations like EDF work with companies broadens everyone's perspective and fosters the sense of working together to protect the environment.

What challenges will the border facilities face by century's end? With the political and economic woes facing Mexico, businesses are fighting for their very survival. Most are not focused on environmental management at this time. Can proactive companies put pressure on other companies? Only by setting an example, because it is difficult to police or influence the policies of other companies. However, if enough companies join together to improve an industrial park, they can make a difference, regardless of outside political and economic forces. Companies with good environmental management systems in place will be among the front-runners in environmental effectiveness and efficiency. They will have reduced their consumption and emission of hazardous and nonhazardous materials, which translates into cost savings as well as environmental protection.

By developing an environmental program that has internal strength and by superimposing more stringent environmental goals than currently are mandated by government, Philips has taken environmental management into a new dimension. The corporation is using it to be the best in environmental care, not only in Mexico, but around the world.

By developing environmental management systems which can withstand economic and political changes, as well as adopting a worldwide environmental standard such as ISO 14000, companies can maintain a standard of environmental responsibility wherever they operate in the world. This is especially true in North America, where there is increasing pressure to adjust environmental standards upward. Companies with environmental management systems in place not only gain economic benefits at the plant level but also can obtain economic benefits by not having many different programs in place in different countries. Thus, standardization in a multinational company's environmental management program can result in significant cost savings overall. This situation now allows environmental management to become a tool for improving a corporation's economic as well as its environmental performance. When these two motivations go hand in hand, more is accomplished in the environmental arena. Economics and the environment: this might be the merger of the millenium.

NOTES

1. SEDUE was created in 1988 and was merged into SEDESOL in 1992. Finally, with the creation of SEMARNAP in 1994, environmental affairs was elevated to ministerial status.
2. See esp. Susan Cohn, *Green at Work* (New York: Island Press, 1995), and Thomas F. P. Sullivan, *The Greening of American Business: Making Bottom-line Sense of Environmental Responsibility* (Rockville, Md.: Government Institutes Inc., 1992).
3. The rationale for adopting common management standards in this area is presented in Stephen Schmidheiny, *Changing Course* (Cambridge, Mass.: MIT Press, 1992).

   Introduced by the International Standards Organization, ISO 14000 is an extension of ISO 9000, which specifies quality initiatives in manufacturing operations, whereas ISO 14000 is dedicated to environmental management systems.
4. See Adam B. Greene, "Seminar on ISO 14000 Brings Together Environmental Managers from Canada, U.S. and Mexico," *Ecoregion* [CEC] 2, no. 3 (Winter, 1996): 3–5.
5. Tony Vindell, "Maquilas Save Money, Environment," *Brownsville (Tex.) Herald*, Sept. 21, 1997.

9

MILAGRO BEANFIELD WAR REVISITED
LOW-LEVEL HAZARDOUS WASTE SITES IN DEL RIO, DRYDEN, AND SPOFFORD, TEXAS

ALFREDO GUTIÉRREZ, JR.

The city of Del Rio, Texas, is located along the Mexican border some 150 miles southwest of San Antonio, 400 miles south of El Paso, and on the Rio Grande across from Ciudad Acuña, its sister city in Coahuila, Mexico. Del Rio's population consists of approximately thirty thousand people, of whom three-quarters are Hispanic. A few of Del Rio's main tourist attractions are Amistad Lake, which bisects the international border; Indian pictographs; rodeos; and visits to Acuña. Nearby Laughlin Air Force Base is the principal source of jobs, in addition to agriculture, food processing, and stores serving both sides of the border.

The original name of San Felipe del Rio was bestowed by Spanish missionaries who arrived on Saint Philip's Day in 1635. That mission was destroyed by Indians, but the name survived until 1883, when the first post office was established and shortened the name to Del Rio to avoid confusion with the town of San Felipe de Austin. However, the name lives on in San Felipe Springs, the primary water supply for Del Rio and, since Indian days, a historic site. The springs are the natural feature most important to our identity, and since the mid-1980s they have been threatened with pollution from several unwanted industrial projects.

The military has been important to Del Rio since the 1850s, when the town was one of a string of cavalry posts protecting the route from San Antonio to San Diego. San Felipe Springs was a designated watering hole for the short-lived camel corps, established by Secretary of War Jefferson Davis. In 1942, the Army Air Corps opened Laughlin Field as a training base for the B-26 light bomber. Deactivated in 1945, it was reopened during the Cold War and was used as a training home for the 4080th Strategic Reconnaissance Wing, a highly secret U-2 unit which obtained the first picture of Soviet missile installations in Cuba during the missile crisis of October, 1961. Laughlin Air Force Base continues as a pilot training base. San Felipe Springs serves it, too, as well as Del Rio. So when a Star Wars project threatened to disrupt the limestone formations underlying this aquifer, for the first time in its history Del Rio drew the line, refusing to accept a military project.

In the summer of 1987, the U.S. Defense Nuclear Agency (DNA) announced plans to construct a Deep Underground High Explosive Test (DUGHEST) site some twenty-five miles north of Del Rio. With a promise of jobs and several million dollars in construction projects, we were informed by the Department of Defense that several underground tests would be held using conventional explosives to simulate a nuclear attack on underground command centers. As these plans unfolded, concerns about threats to the aquifer and the future well-being of the area mushroomed into a public outcry. In July, 1989, buoyed by an outpouring of public sentiment, the Del Rio City Council passed a resolution formally opposing DUGHEST. Armed with a petition containing over twelve thousand signatures opposing the project, I made a trip to Washington with City Manager Jeff Pomeranz to call on our congressional delegation, including Sen. Phil Gramm, Sen. Lloyd Bentsen, and Rep. Albert Bustamante.[1]

Coincident with the "Star Wars" announcement, the city had retained the services of the firm Hogan and Rason to prepare a comprehensive plan for the northeastern portion of the city, including its extraterritorial jurisdiction. The planning area includes some 21.2 square miles, of which about one-quarter is within the city limits. The purpose of this study was to determine the appropriate level and pattern of development, while protecting the San Felipe Springs and San Felipe Creek, a sensitive environmental area, from urban pollution. Prior reports also had studied the threat of pollution, especially from unregulated septic systems. The area's limestone formation, riddled as it is by fissures and caverns, was both sensitive and vulnerable to surface pollution.

Thus we knew it would be inappropriate to allow any development that would threaten both the environment and the integrity of San Felipe Springs. Government officials, however, appeared to know little about these subterranean configurations near the proposed blast site and seemed little concerned with local opinion, or with the public interest. For my part, I had been indifferent to environmental concerns before, but the threat posed by DUGHEST to our pristine limestone springs converted me, like my fellow citizens, into an environmental activist. And when it dawned on us that Del Rio probably had been targeted as one of those places likely to accept the projects nobody else wanted—toxic waste dumps and the like—the stage was set for confrontation.

Also important in our actions was the cooperative spirit that existed with our neighbors across the river in Mexico. Constructed in the 1960s, Amistad Dam is a most visible symbol of the two countries' cooperation through the International Boundary and Water Commission. Each year in the third week of October, a "Fiesta de Amistad" commemorates the agreement between Presidents Dwight D. Eisenhower and Adolfo López Mateos, which they sealed by a symbolic embrace of friendship. Each year this "Abrazo de Amistad" is repeated by local leaders and elected officials on both sides of the border, and the parade is probably the only one which starts in one country and ends in the other, alternating every year.

The upshot is that with this issue, and the subsequent environmental battles we fought, Mexico's involvement made an important contribution to defeating these projects. Mexicans duly protested against the DUGHEST project after being alerted to the possible contamination of our water supply, which came from the same underground aquifer as theirs. It came as welcome news, indeed, when we learned on August 2, 1989, that the Coahuila State Legislature had passed a resolution requesting Mexico's President Carlos Salinas de Gotari to intervene with President George Bush to stop the project. (The Mexican action was welcome news not only in 1989; in subsequent years, too, Mexico protested other potentially damaging projects at Spofford and Dryden, Texas.) Thus we "transborderized" the explosives-testing issue from the start, as well as involving our Texas delegation in Washington.

Our two senators were instrumental in getting DUGHEST not only out of Val Verde County but also out of the state. Bentsen stated that he wanted all testing halted "due to the potential hazard to the region's water resources." When Gramm, on August 29, told the standing-room-only crowd that "there will be no explosives testing in Del Rio," a great cheer went up, and the Del Rio High School Band played.

In welcoming Senator Gramm, I said, "To say that the proposed plan is unpopular in Del Rio is a gross understatement." Furthermore, if the DNA had wanted to do its testing here, "today the citizens of Del Rio were doing the testing, so to speak. Today, we the people are testing whether this nation is indeed a government of the people, by the people, and for the people."[2] Our victory was headlined in the *San Antonio Express-News:* "The people of Del Rio took on a federal agency and won; there's hope for the country yet."

But rejoicing over the defeat of DUGHEST was short-lived, because news soon arrived that big corporations were coming to build toxic waste dumps—three of them, in fact! Citizens groups from three surrounding counties, together with elected officials on both sides of the border, united again to protest three proposed waste dumps in Texas, all within twenty miles of the border. These facilities were:

- The Texcor low-level nuclear waste landfill in Spofford, handling radioactive waste from uranium mining and mill operations throughout the U.S. This would be the only site of its type in the state and one of only two in the nation.
- The Texas Low-Level Waste Disposal Authority's site near Sierra Blanca in Hudspeth County, which would be the repository for spent nuclear fuel from Texas and such states as Maine and Vermont.
- Chemical Waste Management's (ChemWaste) commercial hazardous waste landfill near Dryden in Terrell County, slated to be one of the largest of its kind in the U.S.[3]

We believe that the Del Rio area was picked as part of a "masterplan" by toxic waste companies to target poor, black, Hispanic, Appalachian, or Native American communities for locally undesirable land use projects (LULUs). In 1984, the California Waste Management Board had hired Cerrell and Associates, a consulting firm, to define the profile of communities which were least likely to resist LULUs. Their study drew on a broad range of industry and academic studies, and it was circulated widely throughout the regulatory agencies and in the waste industry. I believe that the Del Rio area was chosen because supposedly it fit the Cerrell profile: rural; conservative; low-income; eager for jobs; following "nature-exploitative occupations" such as farming, ranching, and mining; and not involved in social issues. Del Rio proved them wrong.[4]

We rejected the idea of Del Rio as an area for LULUs and heard of some ways to counter the Cerrell scenario. One of these was to publicize it, which we did, showing how it is the polluters' way to identify "toxic chumps"—the nonpolitical, the poor, the uneducated, all too stupid to resist. Exposing the scenario stirred massive community outrage and stiffened resistance.

As the Texas Water Commission prepared to move ahead with the permitting process, Texas Gov. Ann Richards said she appreciated our concerns but hoped we appreciated that such hazardous waste had to go somewhere.[5] It soon became clear that ours would not at all be the classic "not in my backyard" (NIMBY) defense. No, our concerns were based on genuine alarm about the geological status of our Edwards Aquifer, which was thought to extend well into Mexico. Contamination of shallow groundwater aquifers might feed into the deeper Edwards system which served San Antonio, Uvalde, Del Rio, and areas in Mexico. Low-level and other hazardous materials well might percolate through the formation, as indeed a geological survey of the area (coupled with the first survey ever on the Mexican side) soon confirmed.

We were concerned that contaminated water might be discharged into a nearby creek and thence carried to the Rio Grande and on to the city of Eagle Pass seventy miles downstream. Also, the possibility that radioactive air pollution might be released in the form of both dust and radon gas concerned us. Finally, what would be the long-term liability of taxpayers to clean up eventual problems, especially when the low-level nuclear waste landfills by law became the property of Texas or the federal government?

In addition to concerns about the potential for harm and contamination to our water resources and supplies, we were fearful of the fact that these wastes from all over the U.S. would be transported by railroads and trucks through Del Rio, Sanderson, and Fort Stockton, as well as across the high bridges on Lake Amistad and along routes which were known to have many accidents. It was also pointed out that there were no capabilities in the region to respond to chemical spills, fires, and such.

All these concerns surfaced in the summer of 1988, when the city first heard of the Texcor Industries project. Texcor, before its transformation into a landfill manager, owned the *Comal County Journal*. This Texas company had no prior experience in waste management and no assets except land in Kinney County, yet it proposed to build a four-hundred-acre nuclear landfill in Spofford, just twenty miles from the Rio Grande. The proposed site was to be used for the disposal of uranium mill tailings—that is, byproducts from uranium mining. This proposed landfill was better suited for handling municipal garbage, however, than for low-level nuclear waste. A hearing on the application initially was held before the Texas Department of Health, but the final hearings and decisions were to be handled by the Texas Water Commission. One battle in the war against LULUs had begun.

In September, 1989, Terrell County Commissioner Shirley Spence and environmental activist Marion Childress spoke to the Del Rio City Council on the hazards of another project, ChemWaste's proposed disposal site at Dryden. Slated to be one of the largest commercial hazardous waste sites in the country, this two-thousand-acre facility would accept shipments from the U.S., Mexico, and possibly Canada. If permitted to go forward, it would be the only hazardous waste plant in the border region. The speakers also informed us that, to date, ChemWaste and its parent company had had serious problems with some of their other sites and had been subjected to extremely large fines for violations of environmental regulations or for engaging in unfair business practices. We also were aware that the landfill would be located on porous limestone in an area of many faults and caves. It soon became clear that residents of Val Verde County and the National Park Service (NPS) shared our concern.

Commissioner Spence from Terrell County asked the city council to support the protest efforts with a resolution opposing the project.[6] This the council did on October 12, 1989. The resolution unanimously opposed the development of any kind of toxic or radioactive waste facility at Dryden and directed me as mayor and the city manager to so notify the Texas state government, the state and federal congressional delegations, the appropriate state and federal agencies, and also all the cities drawing water from the Rio Grande, Lake Amistad, the Edwards Aquifer, and our own San Felipe Springs.

In the hard-fought campaign to keep ChemWaste out of Dryden, our main battle cries were "Water Is Life," and "Don't Contaminate Our Water." In fact, contamination of water was our utmost concern, as well as the fear of leakage occurring in the toxic dumps despite their supposedly infallible liners. Spillage into the Rio Grande, a few miles from Dryden, would jeopardize more than a million people on the border, all the way to the Gulf of Mexico. We publicized this danger and invited letters or support or resolutions from all the cities along the border, as well as from the State of Coahuila and its congressman. To local governments along the border, it was emphasized that these toxic dumps violated the spirit and intent of the La Paz Agreement of 1983, as well as the 1944 Water Treaty between the U.S. and Mexico and the Integrated Border Environmental Plan recently proposed by EPA.

At the People's March, whose official name was the "Border Environmental Protest Rally," held on March 21, 1992, a convoy of Mexican citizens from Acuña communities joined with their Del Rio counterparts in a binational demonstration. This binational group halted bridge traffic for two hours and demanded that the Texcor permit to build and operate the Spofford site in Kinney County be denied, along with the proposed Texas Low-Level Radioactive Disposal Authority site near Sierra Blanca in Hudspeth County, and the ChemWaste commercial hazardous waste landfill near Dryden, in Terrell County.

We determined to involve the news media in all aspects of our fight. News releases were issued weekly, sometimes daily. Then, to dramatize our concerns, the above-mentioned People's March was held on March 21. More than six thousand people from both sides of the border met at the middle of the International Bridge between Acuña and Del Rio, stopping traffic for two hours. On the Mexican side, thousands of schoolchildren and concerned citizens rallied in Ciudad Acuña at 11 A.M. On the U.S. side, marchers rallied at the same hour and joined the Mexican protesters at noon. Speakers from both sides of the border, as well as the schoolchildren, were highlights of this unique event, which was covered by local print and radio journalists, as well as television stations from San Antonio, whose helicopters hovered over the march.

We asked for resolutions or correspondence opposing the three landfills, and many cities responded.[7] The NPS also asked to be a party to the hearing process, given the Spofford project's potential for impacting the Amistad National Recreation Area, which draws 1.3 million visitors a year. The NPS was concerned particularly about the impact of fugitive dust coming off the landfill, health risks to humans and wildlife, and the threat of surface and groundwater pollution.[8]

Among numerous letters received from Mexico was a communiqué of April 21, 1992, from Ricardo Mier Ayala, director of the Ecological Department of the State of Coahuila, to the hearing examiner for the Texcor permit at the Texas Water Commission. Mier Ayala referred to a recent Governor's Conference, at which the Texas governor had assured her Mexican counterpart that the State of Coahuila would be granted party status at the Dryden permit hearings, having been granted such status already at the Texas Natural Resource Conservation Commission (TNRCC) hearings for the ChemWaste permit. Congressman Jesús María Ramón, Jr., from Coahuila, and other officials were requesting party status as well. In fact, such was the concern of Coahuila over the Dryden facility that the state had asked Mexico's federal ecology agency, SEDUE, to initiate a consultation process under the La Paz Agreement of 1983. This had begun when officials from EPA and SEDUE met with the Texas Water Commission in February, Ayala added, and he would appreciate a response to this request.[9]

Also noteworthy is the letter from Rodolfo Arizpe Sada, president of the Fundación Ecológica Mexicana, to Susan Rieff, director of environmental policy for the State of Texas. Raising concerns about both the Spofford and Dryden projects, Sada alluded to serious efforts already under way to clean up the Rio Grande. These, along with human health, could be negatively affected if the projects were not relocated to a safer place.[10]

Congressman Ramón was actively involved, along with Dr. Emilio de Hoyos, the mayor of Acuña, our sister city. They asked the Coahuila State Legislature to request that the Mexican federal government take note of the pollution threat to its territory. By now I was actively engaged with all these groups and actors; and this networking, which had been going on for months, soon bore fruit.

Early in 1992, I had made two trips to Washington to confer with our congressional delegation, officials at the Mexican Embassy, and a staff member at EPA. Consequently, on March 2, 1992, the government of Mexico issued the first of four diplomatic notes to the American government, formally requesting information about the intention of Texas to issue permits for two radioactive waste dumps and one hazardous waste landfill on the border. In a statement delivered by the Mexican Embassy in Washington to the U.S. government, Miguel Angel Yunes, president of the Mexican Ecological Commission, expressed the Mexican government's "deep concern for any project which poses a risk to the human safety or natural resources in the Mexican border region."[11] Concerns were raised that the U.S. had violated the La Paz Agreement, or its spirit, by failing to advise the Mexican government in a timely manner of the three hazardous waste sites. In April, an official Mexican delegation, headed by Undersecretary Andrés Rosenthal of Foreign Relations, visited the State Department for consultations on the three Texas waste sites.[12] With Mexico now officially involved, what had begun as a small-town skirmish had escalated almost overnight into an international issue, and this got the attention of official Washington, particularly as this occurred in the months just prior to the U.S. Congress's vote on the North American Free Trade Agreement (NAFTA).

Congressman Ramón already had written several letters to Gov. Ann Richards and to our congressional delegation. Then Ramón and U.S. congressional candidate Henry Bonilla met in Del Rio to sign a joint pledge to fight toxic waste dumps if Bonilla were elected. Ramón also asked that the State of Coahuila and the government of Mexico be granted party status in the Texcor hearing before TNRCC. This was the first time any foreign government had been granted this status in a Texas permitting case, and the Mexican government appeared as a party before the Water Board in the Texcor hearings, too. The involvement of Mexico in these hearings turned out to be a key factor in our eventual defeat of the proposed waste dumps.

On March 2, 1993, an Environmental Awareness Protest March and Rally was held at the state capitol in Austin. Hundreds of marchers from Del Rio, Spofford, Eagle Pass, Brackettville, and Mexico converged on the streets leading to the steps of the capitol. The legislature, in session that morning, was well aware of the demonstration on the steps outside. During a break, several state representatives and capitol employees came out to witness the demonstration.

After the two national anthems had been played, a priest from the Roman Catholic Archdiocese gave the invocation, followed by several speakers from the communities involved. "Why," I asked, "are all three sites located only twenty miles from the border, in areas with low populations, rural lifestyles, and depressed economies? Contrary to the hopes of waste management companies, we are not disenfranchised; the citizens of our communities will stand up and fight to preserve the environment for ourselves, the people of South Texas, and the border communities of Mexico."[13]

By then, citizens groups from all three Texas counties, together with tens of thousands of citizens and many elected officials on both sides of the border, were united in opposition to the three proposed waste dumps. In Kinney County, Madge Belcher was president of the group called Citizens Against Radioactive Environment (CARE). In Terrell County, Lloyd Goldwire headed Sanderson Citizens Against Toxins (SCAT). And in Hudspeth County, Linda Lynch presided over Alert Citizens for Environmental Safety (ACES). The Movimiento Ecologista de Acuña was involved. Ranchers opposing any threat to area water supplies joined the cause.

The Texas Center for Policy Studies in Austin was helping to coordinate what by now had become a binational campaign against the proposed landfills.[14] In the words of Madge Belcher from CARE, "U.S. and Mexican citizens groups are organized to fight to protect all that's important to them—their communities, their water supply, their health, their land, and their livelihoods." She went on to say, "It is incredible for the United States, on the one hand, to expect the government of Mexico to expend over $400 million to clean up the Rio Grande and, on the other hand, to sanction radioactive and hazardous waste dumps close to that same river."[15]

The upshot of all this organizing, marching, and networking was to affect substantially the outcome of the ChemWaste dispute. Del Rio alone spent over $400,000 on legal fees and geological surveys. On May 24, 1995, the TNRCC dismissed without prejudice ChemWaste's request to withdraw its application for a permit for the Dryden project.[16] Other factors played a part. The decision of the Texas General Land Office to conduct a thorough study of the limestone formations underlying the border region—including, for the first time, data from the Mexican side—was important. This study produced results which cast into doubt ChemWaste's contention, based on the company's survey, that the Dryden area was suitable for such a landfill.[17] Raising concern over the waste dumps "also provided the Mexican public and government with an unusual opportunity to direct their environmental concerns toward the United States," the *Washington Post* observed; ordinarily it is the "U.S. environmentalists [who] criticize pollution and lack of government controls in Mexico."[18] Finally, as mentioned, the timing was excellent for activists to raise the hazardous waste disposal issue, during the 1992 consideration of NAFTA and beyond, when the border region took on a high profile.

As of this writing, the Dryden project is dead. Texcor appealed the decision to deny its permit for the Spofford site, and a request to relocate was denied. But on April 5, 1996, the TNRCC approved the Sierra Blanca waste dump for construction. "The fun just never ends," one of the attorneys for the activists wrote after visiting Mexico, where he attempted to get the Mexican government to write yet another diplomatic note. However, this time that government seemed to favor the Sierra Blanca project, and the press was rife with rumors of additional sites under study or in the planning stage.[19]

Let us consider the lessons learned. First, the old maxim, "Strength through unity," never proved truer. The coordinated campaign against the three sites, coupled with our close ties to Ciudad Acuña, played a major role. In fact, the strong bonds of community and friendship that existed with Acuña before were fortified tremendously. Perseverance by the local people and officials on both sides of the border, and the united effort to mount a grassroots campaign, were important in the hard-fought effort.

Information played a role. The realization that we were targeted for LULUs produced a groundswell of anger. The fact that people on the Mexican side were apprised of the possible contamination flowing out of Texas into their groundwater sources was instrumental in getting them involved in our fight. The media played an important supporting role, as mentioned. Timing was important, too, given the sudden prominence of the maquilas (assembly plants) and border environmental problems in the weeks preceding the NAFTA debates in the U.S. Congress. Having these concerns on the front burner definitely worked in our favor.

That we were able to gain the support of two federal governments was also important. In truth, EPA was reluctant to be involved in our early efforts to stop the three projects. Once the Mexican government began sending notes about its concern, however, the projects became a foreign policy issue, which raised the profile of our efforts in Washington. We worked the political system at all levels—local, state, and federal—and binationally, as appropriate. Throughout, we sought solace and justification in being one transborder community united against a mutual enemy.

Above all, we used any and all means of fighting these toxic waste dumps, including letters of support, diplomatic notes, letters and speeches from congressmen and senators, work with the press, rallies, marches, and demonstrations. The campaign was a true partnership, based upon transnational coalition building. In the process, the citizens of Del Rio learned that eternal vigilance is needed to preserve our clean and healthy environment. For example, the fight to stop Sierra Blanca is far from over, and the threat of additional permitting battles hangs over the border.

Del Rio is our heritage, just as our children are our future. Even as we live in an increasingly mobile world, we have learned that this land is on loan to us from future generations. Our responsibility is to leave it in better shape than we found it.

NOTES

1. In all, 12,141 people signed the petition, which was taken to Washington and presented to Vice Adm. John T. Parker, Nuclear Defense Agency director, who had stated emphatically on Aug. 1 that the test would not take place if the community did not want it. However, Parker continued to look for alternative test sites in Val Verde County, choices that still were unacceptable to the citizens of Del Rio. Parker was replaced in early September, and his successor, Maj. Gen. Gerald G. Watson, assured Senator Gramm that, in the face of local pressure, all testing would cease and the facility would not be built in Texas.
2. A cycle of almost daily meetings with the two senators ensued in August. Sen. Lloyd Bentsen to Mayor Gutierrez, Aug. 2, 1989, in collection of Alfredo Gutierrez, Del Rio, Tex.; Gramm and Gutiérrez quoted in the *Del Rio News Herald,* Aug. 29, 1989.
3. The landfill at Dryden also would accept PCBs in low concentrations of less than 500 parts per million. Formed in 1981 to dispose of low-level radioactive waste produced in Texas, the Texas Low-Level Waste Disposal Authority had been looking for a site for several years and recently had lost a lawsuit brought by the City of El Paso, which was concerned about radioactive leaks into its water supply, to block a nearby site at Fort Hancock.
4. Citizens Clearinghouse on Hazardous Waste (CCHW), "Targeting 'Cerrell' Communities," circular distributed Oct., 1995, *Del Rio News Herald* Library, states: "The Cerrell Study is explicit in identifying communities who won't resist LULUs. Because almost every new group (since 1984) served by CCHW matches the Cerrell profile, we believe it's the 'Master Plan' for siting. Cerrell provides important proof that siting is 99% politics and 1% science."
5. Gov. Ann Richards to Mayor Alfredo Gutiérrez, Feb. 26, 1993, Gutierrez collection. "Unfortunately," she wrote, "until all of us do a better job of both minimizing the amount of waste we produce and increasing our re-

cycling efforts, there will continue to be a greater demand for additional disposal facilities, resulting in an increasing number of battles over where to put them. All of us clearly have a massive task ahead of us."

6. Editors' note: Another county commissioner informed the Texas Water Commission that he favored the project, and the Terrell County Chamber of Commerce passed a resolution in favor of the facility, sponsoring a citizens' petition to this effect. Several ranchers in the area opposed it, and the cities of Del Rio, Eagle Pass, and Laredo passed resolutions opposing the facility. Dryden Texas Proposal, Permit Status, Apr. 20, 1992, in City of Del Rio Council Minutes.

7. Del Rio City Council, Del Rio Utilities Commission, Del Rio Chamber of Commerce, San Felipe–Del Rio Consolidated Independent School District, and Val Verde County, all in Del Rio; Devil's River Soil and Water Conservation District; Starr County, in Rio Grande City; City of Eagle Pass; City of Laredo; City of McAllen; City of Brownsville; Well County, in Laredo; City of Uvalde; City of Weslaco; City of Brackettville; and the Association of South Texas Communities (thirty South Texas counties).

8. Regional Director, NPS Southwest Region, to Gordon W. Hardin, Hearing Examiner in Austin, memo received in Del Rio on June 27, 1992, available from Secretary, City of Del Rio.

9. Ricardo Mier Ayala to Gordon Hardin, Hearings Examiner, Texas Water Commission, Apr. 21, 1992, available from Secretary, City of Del Rio.

10. Rodolfo Arizpe Sada to Susan Rieff, Feb. 7, 1992, available from Secretary, City of Del Rio.

11. City of Del Rio, press releases of Feb. 24 and Mar. 2, 1993, available from Secretary, City of Del Rio.

12. Editor's note: The U.S. maintained that considerable information was being made available through the Hazardous Waste Working Group, under Article III of the La Paz Agreement. EPA has been providing information to SEDUE for several years. In February, meetings on Dryden and Sierra Blanca were held among officials of the two federal governments, the State of Texas, and the IBWC; and several other meetings on the three projects took place thereafter. However, what raised the profile of this issue was the diplomatic notes from Mexico requesting specific environmental impact information (particularly concerning crossborder aquifers) and threatening to oppose the three sites. In the March delegation was Victor Lichtinger, future director of NAFTA's CEC.

13. City of Del Rio, press release, Mar. 2, 1993.

14. Richard Lowerre was the attorney who, for the City of Del Rio in cooperation with the Texas Center for Policy Studies, was most involved in coordinating the efforts against Spofford, Dryden, and Sierra Blanca.

15. City of Del Rio, press release, Mar. 2, 1993.

16. Governments and ENGOs who were parties to the hearing included the City of Del Rio, Val Verde Water Control and Improvement District, NPS, Texas Rural Legal Aid Clients' Council, Val Verde County, the Government of Coahuila, the 4th Federal Congressional District of Mexico, the

Ecological Movement of Coahuila, House District 74 (in Mexico), the Texas Rivers Protection Association, SCAT, and ROAD. TNRCC, order dismissing the application ("TNRCC Docket No. 95-0654-IHW; In the Matter of the Application of Chemical Waste Management, Inc. for Permit No. HW-50347-001"), June 6, 1995, City of Del Rio.

17. Editor's note: The Texas General Land Office's involvement turned out to be the first important step in establishing a climate favorable to carrying out the Transnational Resources Inventory Project (TRIP), a major binational geographical information survey of border resources, which is now under way.

18. "Mexico Seeks Halt in U.S. Waste Plan," *Washington Post,* Mar. 22, 1992, A29.

19. Memorandum, Richard Lowerre to CARE, Sept. 27, 1995, available from Secretary, City of Del Rio. David Frederick, of Henry, Lowerre, Johnson, Hess, and Frederick (which represented the Sierra Blanca Legal Defense Fund), to Mayor Gutiérrez, Apr. 26, 1996, available from Secretary, City of Del Rio. Greenpeace Mexico, in press release no. 31, Sept. 5, 1996, reported that, while Foreign Affairs had a long history of opposition to Sierra Blanca, recent studies commissioned by SEMARNAP had been positive. Later, Secretary Julia Carabias of SEMARNAP asked the United States not to approve it (Guillermina Guillen, "Prohíben la construcción del basurero radiactivo en Texas," *El Universal* [Mexico City], July 8, 1998). Like Greenpeace, the Grupo de los Cien opposes the project, which is slated to be the largest of its kind in the U.S. and which already receives sludge from the City of New York. See Lowry McAllen, "Environmentalists Protest Border Dump," *Mexico City News,* May 19, 1996. TCPS *Newsletter,* passim.

10

The Handling of Hazardous Industrial Waste on the U.S.-Mexico Border
A Case Study of TITISA

SERGIO ESTRADA ORIHUELA
AND RICHARD KIY

Across the course of history, the "line" which divides the United States–Mexico border has shifted several times. One realignment resulted from U.S. expansionist aims, which led to Mexico's loss of part of its northern territory in the 1846–48 war. Nature also has been a factor; witness the shifting course of the Rio Grande at the Chamizal along the border at El Paso and Ciudad Juárez. Traditional questions concerning the allocation of surface water recently have spilled over into difficult issues related to the management of groundwater in the shared aquifers that underlie the twin cities along the border. With the advent of plans for hazardous waste disposal facilities, the line between countries has begun to blur, as community groups on both sides of the border have come together to oppose these facilities. In fact, on the question of hazardous waste siting, NIMBYism has become a binational phenomenon. The dictum, "All politics is local," no longer stops at the border.[1]

With community groups from both countries uniting to scrutinize closely the siting of these controversial projects, public concern translates into social pressure and votes. Because the public is so aroused, a project which eight to ten years ago probably would have been approved for construction and licensed to operate, now may be shut down, despite any previous investment which has been made.

Such was the case with Tratamientos Industriales de Tijuana, S.A. (TITISA), a subsidiary of the Illinois-based Chemical Waste Management (ChemWaste). During the early 1990s, ChemWaste spent more than $50 million constructing a hazardous waste incinerator in the Playas de Tijuana suburb of Tijuana, in response to the growing need for waste remediation facilities in that border region. The significance of this investment to the state of Northern Baja California—home to over 750 maquiladoras (assembly plants)—was great; and the need for it, so it seemed, was clear and compelling.

In a 1995 study, Mexico's Instituto Nacional de Ecología (INE) concluded that roughly two out of every three maquiladora plants on the border did not keep accurate records of the hazardous waste generated, and that along the border roughly sixteen thousand metric tons of this material annually was unaccounted for, placing area residents at enormous risk.[2] Among the communities highlighted in the INE study was Tijuana, where ChemWaste back in 1988 had decided to locate its waste incinerator.

ChemWaste's decision to locate its incinerator in Tijuana, the largest Mexican border community, with over 1.5 million inhabitants and over 530 maquilas, made good economic sense. Given the high incidence of illegal dumping of hazardous waste in Mexico and the limited resources to enforce compliance with Mexico's newly enacted environmental laws, at first glance the project also made good environmental sense. In fact, officials at EPA informally had endorsed the project as a first step in responding to the enormous challenges of hazardous waste management on the border.

So what went wrong? Sergio Estrada Orihuela, one of the authors of this chapter, was the regulator charged by SEDUE, the Mexican environmental agency, with canceling ChemWaste's permit to operate the TITISA incinerator. As such, he knows that there are no easy answers. The decision to close the TITISA incinerator, after all, was influenced by several factors, some completely beyond the control of ChemWaste and SEDUE, which originally authorized the permit to build the facility. Rejecting ChemWaste's permit was a painful task, since the project represented one of the single biggest environmental infrastructure investments in Mexico since the passage of the country's foreign investment reforms. Accordingly, the order which Estrada signed had consequences far beyond the community of Playas de Tijuana, where the TITISA facility was sited.

To better understand the circumstances surrounding TITISA, it is important to recognize that, when ChemWaste's original permits were granted in 1989, Mexico only recently had enacted its environmental regulatory statute, the 1988 General Law for Ecological Equilibrium and Protection of the Environment ("Environmental Law"). The new law contained no clear regulations for the siting of waste facilities.

Such guidelines were adopted only recently, in 1996. Accordingly, at the time of TITISA's permitting, Mexican officials did not have a clear roadmap for how to proceed in the siting of such a facility.

Meanwhile, the problems of illegal disposal and handling of hazardous waste were growing and becoming more complex. The Mexican economic crisis of the 1980s lowered wage rates and drew a flood of U.S. and other multinational corporations to the border to establish maquiladora (manufacturing and assembly) operations. In Tijuana, the number of maquiladoras increased from 287 in 1986 to over 530 in just five years.[3] Many of these facilities, such as television manufacturers, electronic assemblers, furniture manufacturers, and metal fabricators, generated hazardous waste. The amount of waste grew, but the Mexican government lacked the resources and the enforcement capabilities to get companies to dispose of their waste properly.

Some foreign investors began to question the transparency of Mexico's governmental permitting procedures and approval process— issues that were supposed to have been addressed under Mexico's new foreign investment law. Even if firms wished to comply, where were they to send their hazardous waste? At the time, not a single hazardous waste disposal site existed on the Mexican side of the two-thousand-mile border zone. Today only one hazardous waste disposal facility with public access is operational in Mexico—ChemWaste's RIMSA facility, located in the town of Mina, Nuevo León. However, that facility is allowed to use only 15 percent of its total site. In the meantime, Mexican national companies are expected to truck their wastes from across the country to this site. A second site at San Luis Potosí has preliminary approval from the Mexican government, but the fate of this project is still pending.

In an attempt at least partially to address the maquila waste problem, EPA and SEDUE in 1986 agreed to an amendment to the La Paz Accord. Annex III called upon U.S. companies operating maquiladoras in the border region to export chemical substances, including hazardous waste, to the country of origin of the raw materials from which these substances were derived. The annex also established notification and consent procedures, which required the country exporting hazardous waste to provide written notice to, and obtain consent from, the country of import prior to commencing export. To date, compliance with Annex III has been spotty at best, and at the time that TITISA's permit was granted, it was largely ineffective.

In his decision to deny TITISA's license, Patricio Chirinos, head of SEDUE, cited among his key concerns the fact that the local population of more than one hundred thousand residents surrounding the plant had changed radically since the license was granted in September 1988. According to Secretary Chirinos, the Playas urban area had expanded to twelve square miles because its population had grown by 4.7 percent annually. These changed conditions, he noted, were "sufficient to revise the terms of the license previously granted." Chirinos, in his April 1, 1992, press conference in Tijuana, also noted that the company had failed to fulfill its license requirements, with many delays and technical omissions, not to mention having submitted a poorly prepared environmental impact study (EIS).[4] Estrada contends that these technical omissions in ChemWaste's permit submission by themselves would have provided sufficient cause to prevent the TITISA facility from operating.

Still, some of the factors with which Secretary Chirinos had to contend were much more political. Among the biggest was the growing political clout of the residents of Playas de Tijuana, a rapidly urbanizing suburb that had become increasingly middle class. In this community, a political action group known as the Playas de Tijuana Housewives Association emerged to oppose TITISA. Its president, Marta Rocha de Díaz, played a pivotal role in mobilizing area residents, as well as in enlisting the support of U.S. environmental groups, such as Greenpeace. In the case of TITISA, one of Greenpeace's key roles was to funnel information on ChemWaste to Rocha de Díaz and her association colleagues. It was Greenpeace which obtained and passed on to the Playas group the information that ChemWaste intended to use the waste incinerator to burn not just maquiladora waste, but also polychlorinated biphanols (PCBs) as well. Alarmed by the health hazards of incinerating PCBs, including the potential release of cancer-causing dioxins and furons, area residents raised a furor.

Another reason Chirinos could no longer remain silent on the TITISA issue was the fact that Northern Baja California was run by the PAN opposition party, which won the governorship in 1988. Tijuana City Hall also was in the hands of the PAN. These basic political facts were not lost on Chirinos, who later went on to become governor of Veracruz. For him, this chance to appease the politically influential community of Playas de Tijuana was not merely a means of improving the environmental quality of a coastal community and protecting residents from potential cancer-causing emissions; it also was a means of scoring some political points for the PRI in PAN territory.

It is worth noting that, at the time of this decision, the state government of Baja California had little power to influence the decision-making process on TITISA or, for that matter, any other environmental project. The decision to nullify TITISA's permit was made by the federal environmental agency, SEDUE. Today this has changed somewhat. As a result of Mexico's new federalist movement, state and local government officials have a greater voice in rejecting projects that could impact their citizens negatively.

In his presentation to the U.S.-Mexico Border Governors Conference in San Diego on April 2, 1992, Chirinos said that "the spirit of the La Paz Accord requires us to act with maximum responsibility to protect the health of our population and to avoid that which can affect the communities on the other side of the border."[5] Chirinos made the point that the decision to close TITISA was an example of binational cooperation, and he urged the U.S. to take similar actions in blocking the Dryden and Sierra Blanca low-level radioactive waste sites on the Texas border (see previous chapter in this book). This connection to the Texas siting issue showed that Chirinos's political agenda was not merely local or domestic in nature, but binational as well.

This is important because the U.S. government at no time opposed the TITISA project, despite the probability that it knew about the impact that incineration of PCBs could have on Tijuana area residents. Furthermore, the environmental impacts of TITISA's operations were hardly transboundary in nature, since air in the San Diego–Tijuana air basin basically moves from north to south. If anything, Tijuana's air pollution problems are attributable partially to pollution from San Diego and not vice versa.

Several U.S. environmental groups, many of them from neighboring San Diego communities, opposed the TITISA facility. Adrian Torres, a local resident of Chula Vista, California, expressed the views of many when he said, "This plant is only five miles from Imperial Beach, but people there don't know a thing about it. The Mexican government is not going to do anything," he added, voicing a common perception. "Our last hope is that San Diego screams about the plant."[6]

During the 1970s and 1980s, the Playas de Tijuana area had become a popular destination for hundreds of retired Americans and Canadians in search of an affordable seaside community. The Housewife Association established ties with this community, mobilizing additional opposition against the TITISA facility.

Concurrently, Tijuana residents discovered, the onset of the NAFTA debates helped draw attention to the border and provided a new binational political context for the siting disputes. U.S. labor and environmental groups were raising serious concerns about the potential negative impact which NAFTA could have on the environment of the U.S.-Mexico border region. It is no exaggeration to say that the border had become the front line of the NAFTA debate. Thanks to NAFTA, many U.S. and international environmental groups who previously had not been active in Mexico suddenly discovered the border.

For their part, Mexican community groups, among them the Housewives Association, now were organizing on the border. In the years since initial approval of TITISA's permit in 1988, then, the political and social dynamics of siting hazardous waste facilities on the border had changed dramatically.

WHAT COULD HAVE BEEN DONE DIFFERENTLY?

Without question, the environmental problems found along the U.S.-Mexico border needed remedies, and they still do. ChemWaste's TITISA was just one of many solutions which could have helped to alleviate them. Unfortunately, the significance of TITISA in addressing the border's hazardous waste management problems never was communicated adequately to the public.

ChemWaste did run a series of full-page ads in local Tijuana newspapers trying to explain exactly what the incinerator was, how it worked, and why company officials thought it was safe.[7] However, ChemWaste did not tell the affected communities about the advantages of the project, nor did it outline measures it would take to safeguard the health of the population from potential environmental risks associated with the existence of waste not properly treated or confined. ChemWaste did not refute the charges made by environmental and community groups specific to their intent to incinerate PCBs. On the contrary, the conclusion reached by ChemWaste officials was that TITISA could be sold as a project of great regional importance. But to the communities surrounding the facility, such arguments mattered very little.

Mexico has no true community-right-to-know legislation, nor is there a requirement to allow for full public comment while environmental impact assessments are being prepared. Such requirements clearly would have helped ChemWaste to assess better the probability of succeeding with the TITISA project. Without question, the availability of a legally required mechanism for public communication or consultation with the general public on questions of environmental impact would have contributed greatly to a more favorable outcome. Or, at a minimum, ChemWaste would have ended up saving time and money.

If there is one thing that can be learned from the TITISA experience, it is that communities need to get involved early in the approval process. Community-right-to-know initiatives, though not fully implemented in Mexico, are an essential ingredient in the environmental decision-making process. Without accurate information, how can citizens hope to make a rational judgment on a proposed project like

TITISA, which may or may not affect them, and where the economic benefits may outweigh the known risks?

Although ChemWaste promoted itself as a world leader in the development of technologies for handling hazardous industrial waste, initially it did not seek a Mexican technical counterpart having professional prestige equivalent to that of its firm. Having the TITISA project evaluated and endorsed by environmental experts from the Universidad Nacional Autónoma de México or the Universidad Autónoma de Baja California could have helped to give the project added credibility, by affirming that the project would have minimal negative environmental impacts or health risks. Still, questions about the incineration of PCBs could have cast lingering doubts.

While the TITISA project in principle complied with all aspects of Mexico's environmental legislation at the time, its application, Estrada says, lacked the detailed technical information that such a project should have had, and its permitting process was handled in a particularly slow and disjointed manner. One could conclude that, if ChemWaste had presented in a precise and timely manner the environmental information required and needed for its evaluation within the framework of Mexican environmental legislation, it is very possible that, on its merits, the TITISA facility would have been authorized in less than a year. Had this been the case, it is unlikely that ChemWaste would have confronted the array of political forces which ultimately led to the project's cancellation. It can be argued, in fact, that TITISA was a victim of timing and of a changing ecopolitical landscape that had becoming increasingly binational in nature.

The most important lesson to be learned from the case of TITISA is that, in Mexico, it is no longer possible to ignore the importance of effective communication and outreach to communities which are likely to be affected by a given project. In the U.S. and Canada this goes without saying, but in Mexico this is still a lesson that is beginning to sink in, only now as controversial projects throughout the country begin to experience the same fate as TITISA. In 1996 alone, the Tepoztlán golf course has been blocked in a highly visible community action campaign; plans for the privatization of PEMEX petrochemical facilities have been stalled, due in part to environmental concerns; and pressure has mounted for Mitsubishi and the government of Mexico to cancel a planned expansion of salt production in Guerrero Negro, the breeding ground of the California Grey Whale.

There is no doubt that, had it not been for community activists uniting on both sides of the border against the ChemWaste facility, TITISA would be in operation today. The lesson to be gained from TITISA, then, is not that ChemWaste erred in not initiating operation of its incinerator more quickly. Rather, it is that, in the siting of such facilities in North America, dialogue between project developers and community groups is absolutely essential. After all, ENGOs, many of them with crossborder links, now are more involved than ever. Yet one should not assume that industry and environmental organizations are implacable enemies. In truth, it is now more important than ever for communities and industry to work together to promote economic development activities that are sustainable. This is another lesson that the case of TITISA teaches us.

It is clear from the fate of TITISA that, to encourage environmental infrastructure investment in projects such as the one that ChemWaste had proposed in Mexico, a country must have well-defined regulatory mechanisms and a commitment to environmental enforcement. Although the politics surrounding NAFTA had a lot to do with the cancellation of the TITISA project, it is fair to say that the growth of the maquila industry, coupled with the prospect of NAFTA and the anticipation of greater enforcement by SEDUE, precipitated ChemWaste's decision to invest in the TITISA incinerator when it did.

Another factor to consider is the impact of U.S. environmental regulations on Mexican industry. Annex III of La Paz requires maquiladora facilities to reexport their waste to the country of origin. Yet, with NAFTA maquiladora benefits being phased out, companies that formerly operated "in-bond," in the future will be treated, legally, more like Mexican corporations. Ultimately this will require that these companies find adequate hazardous waste disposal facilities in Mexico, since Annex III is strictly limited to "in-bond" assembly and manufacturing facilities and not companies deemed to be "Mexican national" operations.[8]

Since SEDUE's 1992 decision to revoke ChemWaste's permit for the TITISA facility, much has changed politically in Mexico with respect to the siting of hazardous waste disposal facilities within national territory. Today such projects undergo much keener scrutiny by community groups and ENGOs, as is clear from the delays currently being experienced by the U.S. firm, Metalclad, in obtaining approval to open the doors of its hazardous waste handling facility in San Luis Potosí. Like TITISA, this site previously had been approved by Mexican environmental authorities (in this case, INE), after an investment of over thirty million U.S. dollars, several years of construction, and environmental impact assessments, including potential risks to the surrounding community. However, because of local and regional opposition to the project's siting and the governor of San Luis Potosí's reluctance to give the green light to this project, the Metalclad facility sits complete, ready to operate, but empty.

Because of lessons learned from TITISA and Metalclad, INE has initiated a comprehensive nationwide effort to identify those locations within the national territory in which the siting of hazardous waste facilities pose minimal or no environmental threat to surrounding communities. In fact, INE recently secured funding from the World Bank to develop a national Geographical Information System atlas which will single out those regions of the country where geological, climatic, and environmental factors are optimal for such sitings. Similarly, INE has formalized a five-year (1996–2000) program minimize, and provide for integrated handling of, hazardous industrial wastes. This program is aimed at facilitating the siting and approval of hazardous waste treatment sites, so as to encourage foreign and domestic companies to invest in such operations. This program includes a mechanism for thoroughly evaluating new treatment technologies before any investment actually occurs in Mexico. Greater public input on any proposed project which could impinge on the community, small or large, also is encouraged.[9]

INE's new approach to defining the terms and conditions under which investment can take place for the development of hazardous waste facilities is driven by the realization that Mexico must solve the issue of hazardous waste siting, despite growing local opposition (NIMBYism) across the country. There now are many abandoned and illegal hazardous waste sites throughout Mexico, and the environmental and health impacts to communities affected by these sites grow annually.

According to INE, Mexico generates an estimated eight million metric tons a year of hazardous waste, with over 63 percent produced by Mexico City and its surrounding states of México, Hidalgo, Querétero, and Guanajuato. Northern Mexico, including Monterrey but excluding the border region, accounts for another 25 percent of total waste generated. Mexico's northern border area accounts for less than 1 percent of waste generated, yet this is the area with the highest rate of industrial growth and the one region of the country where the problems of illegal waste siting and chemical spills become transboundary political issues.[10] Still, at present there exist only two approved sites in the border area where hazardous waste legally can be disposed of.

Beyond the Mexican government's plans to provide a more coherent framework for the siting of hazardous waste facilities, much remains to be done in regard to the community's right to know. While the Mexican Congress recently amended the 1988 Environmental Law, which includes language calling for greater community participation in environmental decision making, the legislation remains silent on the issue of full public access to environmental information on projects which potentially could have a negative impact on the environment. Furthermore, for some industries, the new law weakens requirements concerning submission of an environmental impact statement.[11] Accordingly, Mexico can expect more controversies such as TITISA in the future. Without strong community-right-to-know provisions in its law, this will be inevitable.

TRANSBOUNDARY MOVEMENT OF HAZARDOUS WASTE

Another issue with transboundary implications is emerging, namely the movement of hazardous waste within North America. As we have seen, the La Paz Agreement requires the return of hazardous wastes to the country of origin. The law does permit the importation of hazardous waste materials or wastes for purposes of treatment, recycling, or reuse. EPA and Mexico's environmental agencies (SEDUE, SEDESOL, and now SEMARNAP) have worked jointly to share information, enforce their respective laws, and encourage compliance among border maquiladoras in returning hazardous materials to the U.S. after their intended use. To date, compliance has been minimal. In fact, according to a 1995 report by EPA's Region 9, less than 20 per cent of maquiladora operators along the border of Baja California Norte were returning their waste across the border under the guidelines of the La Paz Agreement's Annex III.[12]

In order to quantify the extent of compliance under Annex III, as well as to estimate better the volume of hazardous waste that is either illegally disposed of or improperly stored along the U.S.-Mexico border, EPA and SEMARNAP have developed a crossborder hazardous waste tracking system called "Haztraks," which registers all crossborder waste shipments on a "cradle-to-grave" basis. The hope is that the information gathered under Haztraks will lead to more enforcement actions and, ultimately, to better compliance with the environmental laws of both countries.

While the issue over the transboundary movement of hazardous waste along the U.S.-Mexico border is well defined—specifically, it involves waste generated by border maquiladoras—the question of whether a crossborder trade in PCB waste should occur across North America remains highly controversial. In Mexico, due to the demise of the TITISA facility (which, as we have seen, sought to incinerate PCBs) and growing opposition to in-country treatment and disposal of such waste streams, INE aggressively has promoted a policy of exporting all of Mexico's inventory of PCB waste. In fact, INE has set forth a goal of eliminating PCB waste completely from Mexican soil by the year 2000. Toward this end, Mexico, as of this writing, has exported 1,669 tons of PCB waste to Finland and England for incineration. Yet the cost of exportation and final destruction of PCB waste in Europe is high, over five U.S. dollars per kilogram. Accordingly, Mexico has promoted the idea of exporting PCB waste to the U.S. and Canada.[13]

Although the politics of crossborder trade in PCBs remains controversial, EPA did authorize the first shipment of twenty tons of PCB waste from Mexico to Port Arthur, Texas, for transport and final destruction at an incineration facility in Oklahoma. This was the first of many future such shipments under a new U.S. policy to open its borders to waste imports in an attempt to assist U.S. waste management companies with excess capacity in their facilities. The Sierra Club Legal Defense Fund has challenged in court the adequacy of EPA's ruling. This case is still pending as of this writing. The Canadian government, too, has provided preliminary approval for the export of Canadian PCBs to the U.S., a reversal on Canada's "closed border" policy concerning PCB transborder shipments.

Responding to the policy shift in the U.S., Canada, and Mexico on transboundary treatment and disposal of PCBs, the CEC released its PCB Management Plan for North America in September 1996. Among other things, the plan advocates the opening of continental borders for transboundary PCB shipments.[14]

Despite the recent push toward opening North America's borders for shipment of PCBs, environmental groups from the three countries, including Greenpeace and the Sierra Club, remain opposed to incineration and crossborder trade in PCBs, due to the potential health risks, including cancer. These groups argue that NAFTA member countries should comply with the spirit of the trade agreement, which calls on each of them to uphold the Basel Convention on the Transboundary Movement of Hazardous Waste. In particular, Basel contains language calling on signatory countries to explore all possible remedies for in-country treatment and disposal of hazardous substances before export is sought as an alternative. These groups contend that the three governments have not done enough to promote alternative environmental technologies—many of which are already commercially viable. They point to the strong political pressures to maintain the status quo and send PCBs and other hazardous organic waste to incinerator facilities in the U.S. and Canada.

Given that PCB is a substance no longer produced or used in industry, its eventual elimination across North America is certain. However, whether this elimination will occur at home, through export to

Europe, or through export among NAFTA member countries remains unclear. What is clear is that the debate over the transboundary impact of hazardous waste movement has not disappeared. In fact, it likely will intensify, as more communities across the continent become sensitive to the public health impacts which could arise from chemical spills along their roadways. Such specific worries will bolster more general NIMBY-type fears about the siting of hazardous waste facilities and incinerators in particular. Despite all such fears, there is a growing recognition—particularly in Mexico—that better hazardous waste management is desperately needed. With this awareness comes the realization that such facilities must be sited somewhere, preferably in places where they will have minimal human impact and where local and ENGO opposition can be accommodated, to the mutual satisfaction of government, industry, and the community.

Without a doubt, questions over hazardous waste siting and the movement of transboundary hazardous waste shipments have intensified in the aftermath of NAFTA. Accordingly, resolution of such questions will necessitate greater cooperation among developers, governmental officials, and ENGOs alike. Successful conflict resolution in regard to such issues will also necessitate greater transparency and public access to information. Without these key elements, the mistakes of the TITISA case are likely to be repeated, not just in Mexico but in the U.S. and Canada as well.

NOTES

1. Much of the reaction to crossborder trade in hazardous waste is well founded. In the last five years, EPA has documented various accidents of Mexican trucks spilling chemical substances on U.S. highways. The AFL-CIO and Teamsters have, of course, capitalized on this concern for their own protectionist purposes.
2. Study was an internal study conducted by the Instituto Nacional de Ecología, 1995; "National Environmental Institute Report Says U.S. Maquiladora Plants Kept Poor Records on Disposal of Waste in 1995," *Excelsior,* Apr. 4, 1996, B2.
3. San Diego Economic Development Corporation, 1995.
4. Francisco Romo and Gregory Gross, "Incinerator License Revoked in Tijuana," *San Diego Union Tribune,* Apr. 2, 1992, p. B3.
5. Ibid.
6. Gregory Gross, "Border Protest Targets Tijuana Toxic Waste Incinerator," *San Diego Union Tribune,* Mar. 22, 1992, B3.
7. Ibid.

8. As mentioned, presently there is only one facility open to the public, and the other four existing dedicated waste landfills are private. U.S. Agency for International Development, *Mexico's Environmental Markets* (Mar., 1995), sec. 3, p. 3. According to INE, in 1992 Mexico generated in excess of 14,500 metric tons of hazardous waste a day, creating an unmet investment need for hazardous waste treatment facilities of 13,325 metric tons a day.
9. SEMARNAP, *Programa para la minimización y manejo integral de residuos industriales peligrosos en México, 1996–2000* (Mexico, D.F.: Instituto Nacional de Ecología, 1996).
10. Ibid., p. 45.
11. John L. Garrison, an environmental legal consultant working for the CEMDA, a Mexican ENGO, in "Environmental Law Revisited, But Will Greater Environmental Protection Follow?" unpublished paper, Nov., 1996.
12. "National Environmental Institute Report Says U.S. Maquiladora Plants Kept Poor Records on Disposal of Waste in 1995," *Excelsior*, Apr. 4, 1996, p. B2.
13. INE, *Plan de acción de PCP* (Mexico, D.F.: INE, Jan. 16, 1996), 93.
14. Commission for Environmental Cooperation, "The status of PCB Management in North America" (Montréal: North American Commission on Environmental Cooperation, 1997), 17.

CONCLUSION

RICHARD KIY AND JOHN D. WIRTH

Citizens across North America have become more environmentally conscious, and their concern has intensified in the face of recent pressures in all three countries to roll back environmental legislation. Moreover, the social base of environmental activism is changing to encompass more locally based groups. Thus, any discussion of environmental management must incorporate the views and voices of a broader public.

To whatever incipient a degree, this public is increasingly trinational in its perceptions and demands. In fact, North America's borders are seedbeds for the institutional and social developments that increasingly will characterize and underpin the continental community. "At their best," it is said, "borders are places where conventional approaches are questioned, stereotypes dissolve in the face of reality, and new understanding emerges."[1] And, as the national barriers to cooperation recede, many new pathways are emerging on both borders, as the case studies reveal.

Polls from the three countries affirm the public's desire for more, not less, environmental protection from their governments. In 1980, only 48 percent of Americans felt that the government was spending too little on environmental protection, but by 1990 this figure had jumped to 71 percent. In Mexico City, 80 percent of residents in a poll in spring of 1996 blamed the federal government for most of the region's environmental problems, citing a lack of resolve among public officials and corruption as the two primary reasons for the problem. And in Canada, recent efforts by the provincial government of Ontario to roll back environmental regulations are broadly unpopular. While it is true that environmental matters never rank at the top of anyone's list of major national concerns—the economy, crime, public safety in general, and illegal immigration (United States); the economy and corruption (Mexico); and the economy and national unity

(Canada), in particular, rank higher—the environment nevertheless is an issue about which citizens feel strongly and about which, according to most, government is not doing enough.[2]

Environmental activism in the 1990s is both less elitist and less dominated by national environmental nongovernmental organizations (ENGOs) than before, becoming more locally based and more community oriented. A good example is the Tijuana Housewives Association, which mobilized its community against a proposed Chemical Waste Management facility, TITISA. Environmental activism also has brought together a broad range of local and transnational groups, something that creates a new political dynamic, as the province of Québec discovered in the dispute over the Great Whale hydroelectric project.

The current environmental activism centers not so much on generic issues such as global warming as on environmental threats that hit closer to home. The NIMBY ("Not in my backyard") phenomenon, once orchestrated by upper- and middle-income groups, now increasingly reflects grassroots activism driven by a new sense of community empowerment. Del Rio's successful fight against the Spofford and Dryden low-level radioactive waste facilities is an excellent example. This small, largely Hispanic community of just over thirty thousand people on the Texas border is among the poorest in America, with over half of all families earning less than twelve thousand dollars a year in combined income. Given Del Rio's limited resources and restricted tax base, the fact that its citizens were able to bend to their wishes such powerful actors as the State of Texas, lobbying interests in the waste management sector, and even the U.S. State Department testifies to the increased impact of citizen-led initiatives throughout North America.

Environmental activism also is characterized by greater networking among ENGOs in the three NAFTA countries, which gives conflict a local/international, or an inside/outside, dynamic. According to the German-based Friedrich Ebert Foundation, of the twenty-nine Mexican ENGOs maintaining active relations with their U.S., Canadian, and European counterparts, all but one received funding from outside the country.[3] Rapid growth of these Mexican ENGOs has been facilitated by technical assistance and help with lobbying. The strategic alliances formed by North American ENGOs have played a role in such high-profile cases as the Paradise Reef pier at Cozumel and the Silva Reservoir bird die-off, which led to investigations by the NAFTA Commission for Environmental Cooperation (CEC). ENGOs intervened in such cases as the planned saltworks project in San Ignacio Lagoon on the Pacific coast of Baja California, and the Metalclad hazardous waste remediation facility at San Luis Potosi. ENGO-led coalitions have affected forest harvesting practices in the Pacific Northwest and in Mexico as well. They are, in sum, a power—at times a trinational power—with which government and business must deal.

THE ROLE OF GOVERNMENT

If environmentalism is the wave of the future, even the most conservative politicians will be turning a shade of green, yet increasingly their postures will revolve around specific environmental issues of concern in a politician's home district and less around ideology. As David Brower notes, "Politicians are like weather vanes in a time when we are searching for compasses."[4] An excellent illustration was support given by New York's Republican Gov. George Pataki to EPA's hard-fought and ultimately successful 1997 effort to strengthen the Clean Air Act. This act addresses the impact of acid rain, ozone, and other pollutants coming into Pataki's state from power plants in the industrial heartland of the Great Lakes region. Among the strongest opponents of this legislation were Democratic congressmen from the Midwest. Rep. David Bonoir of Michigan, who opposed NAFTA and leads the anti–"fast-track" forces in the House, also opposed the appropriation for NADBank, which shifts funds to environmental projects on the southern border.

In another turf battle pitting environmental concerns against regional economic development, Alaska's Sen. Frank Murkowski failed in his last-minute bid to prolong a destructive timber contract in the Tongass National Forest before Congress recessed in October, 1996. To be sure, at his request the Senate did authorize tourboat operators to increase their visits to Alaska's ecologically sensitive Glacier Bay. Ironically enough, Murkowski may have opened an interesting parallel with the Paradise Reef case, in which the alleged failure of Mexico to comply with its own environmental impact legislation has been investigated by the CEC.

A similar trend has been operating in Mexico's Congress. The then-ruling Partido Revolucionario Institucional, on October 15, 1996, amended the country's basic environmental law (the General Law for Ecological Equilibrium and Protection of the Environment) before upcoming congressional elections held on July 6, 1997. This statute for the first time specifies penalties for environmental crimes, while conveying greater powers to enforce environmental laws at the state and local level. It also provides for greater public access to information on projects with potential environmental impacts.[5] However, unlike the counterpart U.S. law, John Garrison writes, "the Law does not grant the public direct access to consult environmental permits or authorizations but rather a summary of such information prepared by government officials, an advancement over the 1988 law but a far cry from true transparency in government. Ultimately public access to environmental information is left to the broad discretion of public officials."

A significant step toward greater public involvement under the new law is that citizens automatically are granted a legal interest (standing) to bring administrative appeals. Previously, ENGOs such as the Grupo de los Cien had to demonstrate a direct interest in an environmental matter before they could appeal a decision by the government. Until then, the chief impediment to success in challenging the government on environmental issues had been systematic denial of legal standing. Moreover, a recent court ruling has bolstered this trend.[6]

The weaknesses that remain in this law are likely to be revisited in the Mexican Congress, now that it is no longer controlled by the PRI. The Mexican Green Party now holds 4 percent of the seats, making it an important swing vote in a legislature where power now is divided among three major parties (PRI, PAN, and PRD).

Linked to the increased political salience of environmental issues is a continental trend to decentralize, referred to as "the new federalism."[7] This is perhaps most obvious in Canada, where the federal government's environmental role is receding, while the ten provinces gain even more authority. To date, only three provinces, Alberta, Québec, and Manitoba, have ratified the environmental side agreement to NAFTA. Yet, in their ways, Mexico City and Washington, too, are ceding more authority to states and municipalities. In Mexico, this trend is driven in part by a recent power shift giving the two opposition parties control of the major urban centers—Mexico City being in the hands of the left-wing PRD, while the other four largest metropolitan areas (Guadalajara, Monterrey, Ciudad Juárez, and Tijuana) are strongholds of the conservative PAN.

In part, the new federalism is budget-driven, as national governments no longer are able to perform all the environmental management functions that the public demands. In part, it reflects some shifting of power away from national capitals toward subnational units, including—at long last—the borders, which no longer are perceived as peripheral areas of nation states, but rather are seen as emerging transnational regions.

This newfound willingness to devolve authority and responsibility favors the growing sense of community along the borders, which is noted in the case studies in this volume. The awareness of ecological interdependence also fosters cooperation. To deal with their polluted airshed, the establishment of the Paso del Norte Air Management

Region permits local actors to participate in formulating subregional solutions to their common problems. As Jamie Alley notes in his chapter, the British Columbia–Washington Environmental Cooperation Council was formed to address common problems, beginning with oil spills and water quality issues. In sum, subnational institutions are emerging to craft, coordinate, and manage similar crossborder problems, but this local problem solving has continental significance because it favors the development of a shared sense of North American place and community.

To be sure, new areas of conflict also are emerging, and this is nowhere more acute than in the management and use of water. The allocation of surface water always has been the main bone of contention between the U.S. and Mexico, particularly with regard to the Colorado River, and both countries have managed the water well, if not entirely equitably. Pressure to renegotiate the 1944 U.S.-Mexico Water Treaty could arise, as the burgeoning growth of southwestern states and Mexico's northern border communities outstrips available supplies. Moreover, worries about groundwater depletion are leading many communities along the Rio Grande to increase their taking of surface water from the river, which is not sustainable.

To the North, the Great Lakes provide perhaps the most successful example of water management between two nation states. But serious issues remain. Great Lakes governments have grappled with implementation disputes arising from the 1967 Supreme Court decision which specified how much water Illinois could take from Lake Michigan for Chicago-area residents. Illinois thought it was in compliance, but improved monitoring techniques recently demonstrated that the flow may average two hundred million gallons a day too much, or about 8 percent more than allowed. Though the amount is but a drop in the Great Lakes bucket, politicians in both Ontario and New York voiced concern that diversions from Lake Michigan could affect the production of electricity as far away as Niagara Falls and the St. Lawrence Seaway. "If Illinois is allowed to divert Lake Michigan for the purposes it believes are in its best interests," a spokesman for the Wisconsin attorney general asked, "what is to prevent, in future years, states like Arizona and Nevada from diverting water for their purposes?"[8]

While grand diversion schemes like the North American Water and Power Project of the 1960s are in disrepute,[9] Canadian resource nationalism still feeds on fears that Mexico and the U.S. at some point will demand access to northern water, to cope with the effects of global warming. (At present, the only way that water flows south commercially is in value-added products such as Clearly Canadian, a bottled water product.) Will the specter of interbasin water transfers to slake the thirst of burgeoning urban areas revive in North America? In the South, traditional disputes over surface water pale by comparison with the emerging conflicts over shared groundwater resources.

Predictions are that the Mesilla and Hueco reservoirs (bolsones) serving the Ciudad Juárez–El Paso metroplex will run dry by the year 2010. Can the cooperative model devised for managing the airshed be replicated in the more difficult case of groundwater, where the jurisdictional tangle is a much more difficult obstacle?

Moreover, the new local-international linkages among ENGOs are trinationalizing some issues, such as the Cozumel case. There the operation of cruise ships at the pier built over the coral at Paradise Reef has put at risk one of the region's and the world's premier reef systems and dive sites. What might have remained a local issue—the threat from increased tour boat landings—is perceived and portrayed in universal terms of reference, as a threat to the common ecosystem.

Sovereignty issues are invoked by polluters who are reluctant to phase out dangerous pesticides in the face of international pressure. Calls to stop the long-range air transport of DDT from Honduras, Guatemala, and Mexico into the Great Lakes evoke the truly continental proportions of some transboundary issues, and complex legal tangles ensue. In the case of Cozumel, developers thought they had complied with existing law, and business groups have reacted negatively to the CEC investigation. But citizens and groups appealing to universal values are challenging this defense, especially when a resource is defined as being important for the common good. This recently was demonstrated in Holland, where courts ruled in favor of ENGOs which argued that the habitat of pelicans who live in coastal waters is a common ecological resource that would be put at risk if Shell Oil's plan to expand offshore oil production went forward.

While the battle on environmental issues deemed to be extraterritorial has just begun, governments are having to deal with new combinations of actors, as Mexico learned in the case of Cozumel. Lacking effective access to national courts, Mexican groups and their allies found in the CEC a way to force the government to respond to their concerns.[10] That environmental groups had to go outside their own country to push legal action against environmental law violations is striking, because Canadians and Americans already have this right. (As noted above, Mexican citizens recently were granted the right to bring administrative appeals.) But Cozumel cannot be viewed solely in national terms: it effectively was defined from the start as a transnational issue.

For its part, the Mexican government is responding to growing environmental activism by setting up a public consultation process before permitting and siting battles erupt. This is being tried with the proposed San Ignacio Saltworks project for the Sea of Cortez. The advisory panel, composed of academics, business representatives, ENGOs, and government officials, seeks to create a consensus on policy decisions before construction begins.[11] This brings Mexican practice into line with Canadian and American procedures, whereby the management of conflict is facilitated by citizen access to information, transparency, and public comment. However, a single example of this process could still be seen as an exception to the rule, since no such process is mandated by national legislation.

Familiarity with public participation should ease acceptance. As David Cliche relates, the Québec government came to embrace public consultations as the best way to defuse conflict over energy issues and then encouraged public debate in order to build a constituency for its new energy policy. Making information available to hitherto excluded groups does complicate the policy process, to be sure, but for all the right reasons. Thus, the arrival of pluralism in the Mexican Congress is liberating, for the first time, information on the financing and use of water. This encourages localities far from the Federal District (D.F.) to protest the diversion of their streams and rivers to slake the thirst of "chilangos," and the sending of raw sewage from the D.F. back down other rivers to the Gulf of Mexico.

The responses of governments in North America to citizens' demands for practical solutions to environmental problems remain fragmented and uncoordinated, as Roberto Sánchez and his colleagues reveal in their survey of existing transborder programs. Particularly striking is the fact that trinational programs still are few and far between. Given the popular mandate for increased environmental protection, the public policy implications are clear: more coordination among governments at all levels is required, more public consultation and consensus building are necessities, and more cooperation to advance a regional, or continental, agenda is needed.

BEYOND NAFTA

Although this book has focused on the interior borders of North America, the discussion of conflict resolution and cooperation is relevant to the evolving pattern of trade and environmental relations between the two major trading blocs, NAFTA and Mercosur. Once physically isolated due to poor infrastructure, restrictive trade policies, and general underdevelopment in its border regions, South America has faced few transboundary environmental issues to date. This situation is changing, as economic growth and expanded crossborder trade bring with them such transboundary environmental problems as air and water pollution, flooding, hazardous waste spills, and the destruction of forests with consequent desertification and soil loss. Over eighteen thousand miles of frontiers are shared by South American nations—ten times the length of the U.S.-Mexico border.[12] Given that continent's huge need for infrastructure, Brazil and other countries are pushing ambitious development programs, and the demand for capital is estimated to exceed sixty billion U.S. dollars a year. With highways, dams, and pipelines on the drawing boards, the investment climate is sure to be affected as citizens and ENGOs mobilize to confront the environmental consequences of rapid integration under Mercosur.

Although South American ENGOs still are relatively weak in comparison to their North American counterparts, the movement there, too, no longer is limited to the upper and middle classes, but is reaching peasant groups and a host of local actors alarmed by threats to their communities. According to the Mexican political scientist Victor Toledo, environmentalism is catching on among the peasants in South America, and he contends that a major backlash could occur against projects that directly threaten them.[13] This trend will affect governments and development interests, where large infrastructure projects destructive to the environment, such as mining projects, oil exploration and production, gas pipelines, logging, toll roads, and dams, are being taken on by grassroots groups allied with national and international ENGOs.

As a case in point, on March 15, 1997, the Chilean Supreme Court ruled against a major logging project which had been expected to bring sixty billion dollars of foreign investment to Chile. This ruling against U.S.-based Trillium Corporation shook the Chilean forestry industry, known for its political clout. It also took President Eduardo Frei by surprise, since he had spent three years lobbying for this project, which would have resulted in the logging of 257,000 hectares of Lenga forests, a species unique to southern Chile. What happened to Trillium is similar to some of the cases in our book. While Trillium voluntarily submitted an environmental impact statement (EIS) in conformity with the country's newly approved environmental legislation, not all of the law had been enacted, and crucial sections of the EIS process had not yet been defined. It happened that the technical agency of the Regional Environmental Commission (COREMA) rejected Trillium's claim that its logging practices would not ruin the forest, but COREMA itself, being largely staffed with political appointees, approved the project. Environmental activists teamed up with Congressman Guido Gurardi to mount a legal challenge. Until a full regulatory regime was in place, the court held, COREMA had no authority to approve EIS studies and give the go-ahead on such projects. Worried about the chilling effect this case could have on foreign investment, the government rushed through the required regulations, and the whole EIS system is, as of April 3, 1997, legal and compulsory.[14]

Most recently, in February, 1998, the Chilean Supreme Court overturned its earlier decision, on grounds that Trillium Corporation was working closely with community environmental groups and indigenous people to minimize the environmental impact on Tierra del Fuego. Dubbed Proyecto Condor, Trillium's response to regional environmental concerns has yet to convince some ENGOs.

Grassroots actions to oppose such megaprojects as the Great Whale project in Québec and the nuclear research station on Lake Pátzcuaro, now are being replicated further south, most notably in opposition to dam projects such as Argentina's Corpus Dam on the Paraná River. In a recent nonbinding referendum, residents of Misiones Province voted nine to one against this long-planned 3,000-megawatt project—scarcely an encouragement to potential investors for a four-billion-dollar (U.S.) project which is slated to be funded entirely with private capital. Hidrovía, a barge canal project on the Paraguay River which would bisect and dredge sensitive wetlands in Brazil's vast Pantanal bioreserve to service soybean producers, has been redesigned in the face of ENGO pressures and international concerns.[15] Megaprojects of this sort well may suffer the same fate as Québec's Great Whale project, once the environmental cost accounting is done and the inside/outside dynamic comes into play.

Projects that local groups want, but which are opposed by others (including foreign ENGOs) on environmental grounds, will raise complex and troubling North-South issues. A case in point is the projected highway paving project across the forest of northern Amazonia, from Manaus, Brazil, to the Venezuelan border. "Build a road and people will follow it" is a maxim among anthropologists and others concerned with the predations of squatters in the rainforest. But farmers, miners, and ranchers want this road, and government proponents of the project point to the advantages of a land link between two major countries which historically have been distant in their rivalry over oil and regional prestige.[16] When conservation clashes with local interests, the struggle to find pragmatic solutions depends upon one's analysis of the seriousness of the environmental threat. Now that Asian loggers are licensed to operate in Brazil, the prospect of rapid destruction of the Amazon high forest is certain to raise similar North-South concerns about a vast region already suffering from the impact of squatters burning the land and miners polluting the streams.

Environmental conflicts over natural resource issues are proliferating, such as, for example, oil drilling on the Venezuela-Colombia frontier, and mining and logging in the Venezuelan Amazon which threatens the Imataca Natural Reserve. While Venezuela's Ministry of Environment and Natural Resources has ruled in favor of the Imataca project and asserts that opposition to it has been manipulated by "false" environmentalists, there is no guarantee that the concessions being granted to foreign mining interests (including U.S. and Canadian companies) will stand up in court. The political winds are changing in Venezuela. And when a country's environmental regulations still are not fully formed, projects can be stopped, as happened with the Mexican hazardous waste projects. Rather than accepting the assurances of politicians, foreign companies would be well advised to follow the best environmental practices accepted internationally.

When local groups protest environmental damage, as happened recently when Venezuelan fishermen blocked the main Maracaibo shipping channel to protest pollution from a grounded tanker, it well may be that the time has come for regional stakeholders to discuss issues arising from development and the environment. Examples of how this could be done are provided by the Gulf of Maine Council, and the Business Council for Sustainable Development of the Gulf of Mexico. Might not a similar organization be formed for the Lake Maracaibo region?[17]

This is not to say that regional environmental cooperation is absent in Latin America. Several examples have been surfacing. When federal officials refused to act, state and local officials in the mountainous Sierra de Perija on the Venezuela-Colombia frontier stepped in to address common issues such as drug running and water contamination.[18] Another example is the Argentine-Chilean Joint Commission, established jointly to manage the natural gas fields and fisheries of the Beagle Channel, once a potential war zone between these traditional naval rivals. The two nations have also agreed to cooperate on mining concessions and glaciers along their borders.

Transregional trade and environment issues also are coming into focus in recent U.S. court cases. Unable to gain redress in their own courts for damages caused by Texaco and other oil companies to the Ecuadorian rainforest, community groups, including Indians and settlers, brought suit in a federal district court in New York, where Texaco is headquartered. The government of Ecuador asked to be recognized as a party in this case, in support of the plaintiffs. As of May, 1998, the case was still pending.

Charging damages to the health and environment of their city by the Southern Peru Copper Corporation, a U.S. copper mining and smelting operation, the Committee for the Defense of the Health of Ilo brought suit in a federal district court in Texas, over the objections of Peru, which backed the company. In 1995, this smelter was emitting two thousand tons a day of sulphur dioxide into the air, "causing smoke so thick it hovers over the city like a fog and sends residents to hospitals, coughing, wheezing and vomiting."

Judges in both courts rejected the suits on grounds of *forum non conveniens*—that is, the suits were not proper for U.S. court jurisdiction, because the relevant incidents occurred in another country.[19] Whatever the ultimate result, these suits point up the need for new ways to deal with transnational pollution. One happy immediate result: unfavorable publicity generated by this suit prompted ASARCO, the majority owner, to install controls on the Ilo smelter.

When a company engages in industrial practices abroad which are no longer allowed in the country of domicile, how and when does the polluter pay? This is sure to be an issue in North-South environmental relations.

The lack of a trade-environment linkage in the Mercosur Treaty must be addressed if the Mercosur and NAFTA blocs are to cooperate and then harmonize their statutes and regulations. Mercosur faces such important challenges as incorporating the public participation process (which is not now even on the agenda) into its structure, the strengthening of environmental concerns within each of its member countries, and the management of their shared ecosystems. The participation process is incorporated into NAFTA, even if it is not yet incorporated fully into the political cultures of all three countries.

Opposition to linking trade and environmental issues is still strong in South America. President Samper of Colombia, at a meeting of the Non-Aligned Movement in Jakarta, proclaimed that "Colombia and other Latin American countries support Asia's view that human rights, environment, and labor standards should not be discussed at the World Trade Organization."[20] On the other hand, to secure a trade advantage in the U.S. market over petroleum from Arab countries, Colombia, Mexico, and Venezuela are discussing the use of environmentally sustainable exploration and production practices and environment-friendly technologies in their respective oil industries.[21]

Just what weight the so-called "social issues" of environment, labor rights, and democratic governance will play in the coming interface between the two trading blocs remains to be seen. Certainly the "Spirit of Miami" evinced in late 1994 was promising, but the failure of the U.S. Congress to deal with Chile's bid to join NAFTA with the side agreements is disheartening. Chile—having signed bilateral trade agreements with Mexico and Canada (the latter with side agreements modeled on the NAFTA accords) and given its new status as an associate member of Mercosur—could be a bridge between the North, which has embraced the trade-environment paradigm, and the South, which resists it. As of this writing, the policies of Brazil and the U.S. still do not agree about linking trade with environmental and labor side agreements in an expanded accord. Although the two nations are coordinating their positions on the environment at international conferences, and the bilateral agenda was advanced by President Clinton's recent state visit to Brazil in October, 1997, the U.S. and Brazil have not reached an agreement on climate change.[22]

Will regional ENGOs, supported and abetted by their foreign counterparts, force these countries to include green provisions in any agreement expanding Mercosur? The inside/outside linkage among ENGOs, so important to North American activism, is still weak. Some eight hundred environmental ENGOs are registered in Brazil alone, but the potential of these fledgling organizations is still latent. And it remains to be seen how the World Trade Organization, with its new trade dispute panels, will affect relations between NAFTA and Mercosur in general, and will shape the trade-environment linkage in particular. Whether NAFTA can export its innovative and effective model for the role of social issues in trade treaties is very much an open question.

Will President Clinton be able to keep the "Spirit of Miami" alive in his second term? Certainly the administration's battle over "fast-track" is going poorly, with Democratic opposition to free trade now rampant and with Republicans opposing the granting of such authority with environmental and labor side agreements. Trade and environmental issues are at a critical turning point. The CEC, an institution spawned by NAFTA and the first operating agency in a trade agreement to deal effectively and equitably with environmental issues on the North-South agenda, someday may develop "legs" to South America. Meantime, the slowing of momentum for a Free Trade Agreement of the Americas does give this innovative institution, which is generating good product and adding value, breathing room to consolidate its role in the North.

THE ROLE OF BUSINESS

Having highlighted the role of governments and some of the institutions of civil society, we note that business also has a central role to play in the management of transboundary environmental issues. Many of the conflicts profiled in this book could have been averted or minimized if the corporate officers involved had been more sensitive to the effects of their projects on local communities, who then felt compelled to react, and to the ENGOs, which should have been dealt with early on. The Trillium story is a case in point: by becoming proactive, they were able to reverse a court ruling and procede with their forestry management project. Positive efforts on the part of businesses are also profiled in the chapters by Ann Pizzorusso and by Pete Emerson and his colleagues.

An important and central feature of many conflicts was a lack of publicly available information on a project's likely environmental impacts. In some cases—Sierra Blanca, it is being argued—economic gains outweighed the environmental costs. In other cases—Carbón II, according to its supporters—siting coal burning in a remote area outweighed the damage to visibility in Big Bend National Park. Even when the environmental impact is minimal, without publicly accessible information, a community is forced to rely on rumor, speculation, and innuendo, and the result can be delay, defeat, and a souring of the investment climate.

Although Mexico has yet to implement full community-right-to-know legislation, such as exists in Canada and the U.S., companies would be well advised to undertake their own community outreach campaigns during planning and well before construction. They should provide as much information as is practical or necessary to allay community fears and apprehensions.[23] Many Mexican business and industrial interests oppose such legislation, but in the recently revised environmental law, the Mexican Congress has opened up new space on the question of the community's right to know. Voluntary outreach efforts cannot guarantee that a given project will go forward, but they can minimize the chance that a company, proceeding simply on the basis of having secured government approval, in the end will be blocked by public opposition. TITISA is a case in point. Another good example is the opposition by environmentalists to the plans of Metalclad, a California-based company, to construct a government-approved hazardous waste dump in San Luis Potosi. The upshot of these two cases is that not a single hazardous waste containment project has gone forward in Mexico since 1993, the year NAFTA passed, even though the waste piles up and must go somewhere. Metalclad and TITISA are examples of projects that were thwarted well into the construction phase. Had their respective developers been more sensitive to the potential environmental vulnerabilities, measures could have been taken to minimize them or simply to select an alternative project site, where environmental risks were not as great, avoiding large losses. In two other cases, Molten Metal Technology, on its own initiative, responded to concerns about its plans to build a recycling plant for toxic materials including cancer-causing PCBs in Coatzacoalcos,

Veracruz, but was rebuffed. In 1996, a contract was given to RIMSA to export some of the PEMEX, PCB waste to Texas, which the Sierra Club then challenged in court.[24]

The opportunity to play a positive leadership role in reaching compliance with standards mandated by law, while benefiting the environment, has opened up through the new emphasis on joint implementation between business and government, which can even involve ENGOs. The ISO 14000 training program on the border, profiled by Ann Pizzorusso, is another important development. In Mexico, encouraging small- and medium-sized business sectors to adopt better environmental management practices is perhaps the greatest single challenge to business leadership with respect to the environment. Will such corporate leadership efforts be in vain? Given the high costs associated with becoming certified and coming into compliance with ISO 14000, such environment management programs, while good in principle, may have limits. Even U.S. companies have been slow to take up the ISO 14000 flag, in contrast to their European counterparts. However, this could change, now that EPA is considering incorporating ISO 14000 as a qualification criteria for participating in some of its voluntary compliance initiatives. The standard is being considered for adoption by the Department of Justice and the U.S. Sentencing Guidelines Commission in conjunction with enforcement policies and directives.[25] Management certification under ISO 14000 might even become a new form of "green labeling" if consumers demand it.

Partnerships are "in." The Ciudad Juárez brickmakers project is an excellent example, even if it is a cautionary tale of the severe difficulties faced by groups seeking to encourage change in the informal sector. Great Lakes United is the largest and best-developed organization of its kind, even if it now may suffer the effects of federal cutbacks. Faced with a looming environmental crisis in the Gulf of Mexico, industry, tourism, and fishery interests found that they shared a common ecosystem and common ground. This cooperative spirit gave rise to the Business Council for Sustainable Development in the Gulf.[26] A similar institution for the U.S.-Mexico border is long overdue. Meanwhile, Texas state agencies are promoting partnership solutions to transborder problems with a range of private and public groups in Mexico and the U.S.[27]

Chemical industry associations have established the Responsible Care program among member companies across the continent. Aimed at promoting pollution prevention and environmentally responsible business practices, this program now has been adopted by many chemical manufacturers on the southern border. Similarly, some U.S. companies voluntarily have adopted EPA's 33/50 emissions-reduction program there. However, for many years, U.S. parent company owners of maquiladoras kept a low profile, shunning media attention for fear of raising the specter of the loss of American jobs to Mexico. This was particularly true of the Big Three auto makers and their parts suppliers, which had strong labor union constituencies back home. Suspicious of organized labor, maquiladoras ignored criticism of their operations, including their environmental practices. This policy backfired during the 1993 congressional debates over NAFTA, where the border's image as a "pollution haven" became fixed in the public eye.

The irony was that companies which were doing a good job of responsible environmental citizenship got no credit. Because of industry's general reluctance to open its doors to community groups and ENGOs, the image of maquiladoras as a whole became tarnished, as a result of the misdeeds of a few. Meanwhile, efforts to promote environmental management proceeded, particularly among the bigger plants, particularly Sony, Allied Signal, and Philips. Much remains to be done; better information policies are needed, and open doors are required. Otherwise, the image of massive industrial pollution will persist, despite the fact that the biggest contributors of border pollution are untreated residential sewage and air pollution, mostly from cars and unpaved roads.

Because changes in the sociopolitical landscape along the internal borders are so dramatic, the consequences to a company of a mishandled project can be quite costly. It therefore is important to recognize that "environmental consciousness," and the widening social base supporting it, must not simply be dealt with, but forthrightly accommodated.

LEADS FOR FUTURE RESEARCH

Without question, NAFTA and the intensifying growth of highway and rail corridors among Canada, the U.S., and Mexico have turned a spotlight on an array of transboundary environmental issues which hitherto had received minimal, or at best sporadic, attention by government, industry, the media, and even ENGOs. Dealing with regional environmental problems has resulted in various conflicts but also has served as the catalyst for cooperative efforts among diverse interests from across the continent. Furthermore, the environmental challenges facing North America and the rest of the Western Hemisphere are certain to intensify. The case studies in this book, while representative, are hardly definitive. Rather, we hope they can be a starting point for studying the North American environmental agenda. Dissertations, books, and articles yet to be written well may deal with the following issues.

The Maritime Borders

Though often overlooked, these borders are where some of the most complex challenges will be faced with regard to environmental and natural resource allocation. For example, dwindling fish stocks in the Gulf of Maine result from mismanagement, overfishing, and point-source pollution from the growing urban communities that surround the shores of this binational ecosystem.[28]

Similar challenges confront the Gulf of Mexico, ranging from urban and agricultural runoff to sewage and hazardous waste discharges, marine debris, oil spills, destruction of its fertile wetlands, and the impending extinction of a variety of endangered species. Only cooperation among the region's various industries, including fisheries, tourism, oil, petrochemical, and chemical, can reverse the various biological "dead zones" that now exist along the Louisiana coast and threaten the fragile estuary of this ecosystem.

At the confluence of Alaska, Yukon Territory, and British Columbia, the Tatshenshini wilderness was threatened by the Windy Craggy project, a massive open-pit copper mine that could have polluted the watershed into Glacier Bay, a U.S. National Park since 1980. A successful binational effort to safeguard the Tatshenshini culminated in 1993 when the one-million-hectare region of northern B.C. was declared a Provincial Park. Researchers interested in clashes arising from the often-conflicting goals of resource development and wilderness or habitat preservation, and the traditional lifestyles of native peoples, should pay close attention to this classic case of cooperation across jurisdictions touching on another of North America's maritime borders.[29]

On the U.S.-Canada Arctic border, the impact of oil and gas exploration in and around the Beaufort Sea is pitting the State of Alaska and private sector oil interests (including some landowning Native Indian Corporations) against the Inupiat of Alaska, who fear that oil and gas production would have an adverse effect on their marine life, especially on the bowhead whale, upon which they depend for food.[30]

Native Groups

As has been shown in the cases of Hydro Québec and the Beaufort Sea, Native American populations are exerting their political rights to manage growth and environmental problems in their traditional homelands. Issues are framed within the context of complex treaty and sovereignty rights. The view is different if one comes from Nunavik, the homeland of the eastern Inuits, than if one comes from an energy-exporting province encompassing much of the same territory, called Québec.

On the other hand, in 1992, EPA had to deal with complaints from Mexico and the State of California over plans by the Campo Indian Reservation to build a solid waste treatment facility located above a sensitive aquifer within miles of the border.

Pan-Indian movements are becoming important. In 1977, the Inuit Circumpolar Council (ICC) brought together one people living in four separate countries. Since its inception in 1992, the eight-nation Arctic Council has included the participation of native peoples. The principal goals of the ICC are:

- To strengthen unity among the Inuit in the region around the North Pole.
- To promote Inuit rights and interests on an international level.
- To develop and encourage long-term policies which safeguard the Arctic environment.
- To seek full and active participation in the political, economic, and social development of the circumpolar regions.[31]

Now disrupted by political unrest, Pan-Mayan activism on the Mexico-Guatemala border is expected to grow, with cultural survival and access to land leading the demands.

Environmental Justice

This is another area that warrants more attention from researchers interested in understanding better the changing political landscape surrounding environmental issues in North America. Mayor Gutiérrez's account of the City of Del Rio's battle against the Spofford low-level radioactive waste site raises the issue. Failure to understand the potential impact which a project can have on minority and economically disadvantaged communities can lead to its cancellation.

This painful lesson British Nuclear Fuels (BNF) learned as its $750 million uranium enrichment project in Claiborne Parish, Louisiana, came to a halt after the U.S. Nuclear Regulatory Commission denied an operating license on grounds that racism may have played a part in its decision to site the processing plant on the doorsteps of two black communities. Known as "CAN'T," for Citizens Against Nuclear Trash, this community-based action committee took advantage of a 1994 executive order signed by President Clinton requiring that all federal agencies protect minority communities from "environmental racism." As one resident noted, "This is not a case of NIMBY, it is a textbook case of PIBBY—place in black back yards."[32] More such rulings can be expected along the U.S-Mexico border, as Hispanic communities reject unwanted waste dumps, sludge plants, landfills, and incinerators.

Native Americans in the U.S. and Canada long have confronted such Faustian bargains in their struggles to escape poverty without ruining their environment.

Trade Corridors

The impacts of NAFTA-generated trade only now are being felt along trade corridors that are under construction, being upgraded, or being planned. Each of these projects raises environmental issues, some more serious than others. The so-called "NAFTA Highway" from Montréal to Mexico City has seen the volume of truck traffic increase by 25 percent a year, as measured at some border crossing. Meanwhile, railroad trackage has been purchased and leased to create the first truly North-South railroad network in North America, which is ideal railroad territory. However, tracks still run through the center of many small towns and cities, and dispatchers have not yet learned how to handle a backlog of cars on the southern border. Particulate pollution from diesel engines is increasing rapidly. Locational decisions attendant upon the increased road and rail traffic must be made. The dispute over road safety standards continues between Mexican truckers and the Teamsters, but both groups are protectionist when it comes to access to their own highways. The lack of a North American transportation plan soon will be felt acutely, and residents of border cities are upset that governments have done little to deal with an inevitable result of growth—pollution. Why should they bear the costs of increased trade?

Port facilities at Searsport, Maine, are being upgraded and expanded in conjunction with improvements on the TransCanada Highway. The effects on the Gulf of Maine resulting from pollution runoffs into the St. Croix and St. John's rivers bear watching.

The projected intercoastal waterway across the Laguna Madre Estuary, and the interests and resources at stake, should be documented in full. The State of Tamaulipas is vigorously promoting this link to the Gulf Intercoastal Waterway to expedite barge traffic from the port of Tampico to Corpus Christi.

A project with similar environmental impacts is on the drawing board for the Gulf of California Delta region, as binational economic development interests in Mexicali, on the Mexican side; Calexico, California; and Yuma, Arizona, promote the development of a binational canal leading from the Sea of Cortez to the landlocked hinterland of Northern Baja California and Imperial and Yuma counties.

Transboundary Air Pollution

The acid rain debates of a decade ago should have alerted researchers to the interconnected nature of the North American airshed, yet data still is being collected along national lines, using different standards and equipment. *Continental Pollutant Pathways*, the CEC's landmark study, draws attention to these deficiencies, as well as showing the trinational patterns of exchange. This way of thinking needs to be applied to specific industries, as, for example, the effects of deregulated natural gas and power line transmissions on various regions of North America.

More research is needed on energy conservation.

The loading of DDT and other harmful pesticides from Mexico and Central America onto the Great Lakes is a wakeup call for continental action.

Plans for the Carbón II power complex did not trigger outcries by the watchdog ENGOs on both sides of the border, which points up the down side of such activism: they are stretched thin and cannot do everything. Moreover, they are responsive to different constituencies but often compete for the same pool of funds. If the three nations had agreed to a Transboundary Environmental Impact Assessment protocol, these plants could not have been built without controls.

With respect to compliance with the Clean Air Act, El Paso, acting alone, cannot meet the national standards. Only by treating the Paso del Norte as a transboundary airshed can air quality issues in the Ciudad Juárez–El Paso region be addressed. Similarly, New England cannot comply with the Clean Air Act unless pollutant pathways bringing NO_x and SO_x from Toronto and Québec are abated in a coordinated regional approach. Both these regions themselves are impacted by air pollution coming from the American Midwest. This is a set of practical problems worthy of concerted action by the three federal governments, based on good scientific and economic analysis.

Emergency Planning and Response

Communities along the Red River from Wahpeton, North Dakota, to Lake Winnipeg in southern Manitoba together experienced an overwhelming national disaster. This "Flood of the Century" in 1997 united community groups from both sides of the border in efforts to minimize the damage and loss of life. The Internet provided accurate, real-time news reporting and basic information on emergency procedures and how to minimize water damage to family heirlooms and bring back flooded lands and farmsteads. Victims flooded out of their homes were housed quickly and efficiently across the transborder region. Much can be learned from the way Red River residents took advantage of the advances in information technology to promote an effective binational emergency response and disaster preparedness network.[33]

In fact, border communities across two borders participate in emergency planning and response exercises. Notable among them are the Brownsville–Matamoros agreement on contingency planning and the San Diego–Tijuana Emergency Planning and Response Committee. Eventually, fourteen such plans will be developed by paired cities along the border, with assistance from the binational Work Group on Contingency Planning and Emergency Response, established under the La Paz Agreement.

Joint public health planning to monitor and contain contagious diseases is well under way. However, the spillover from these activities into other patterns of cooperation runs up against a basic reality: national capitals, preoccupied as they have to be with drugs and migration, are slow to recognize that the accelerating networks of social and economic links on the borders are running well ahead of our political institutions.

The Role of Leadership and Contingency

In assessing what works or doesn't work in crossborder management, the emergence of a highly placed champion can be critical. The willingness of SEDESOL's Luís Donaldo Colosio to bless the Paso Del Norte transborder task force's approach to regional air pollution abatement is a case in point. Colosio eased the way for official Mexico to accept this highly unusual local initiative on the sensitive Northern border, after which the U.S. State Department eventually acquiesced. Similarly, the friendship between a Washington State governor and his British Columbia counterpart was critical to the early progress of their joint Cooperation Council. But starting in 1997, the next B.C. premier put Council meetings on hold because of the salmon fishery dispute, although the task forces continued operating. In the one case, the actions of a champion interested in addressing border problems redefined the traditional sovereignty issues; in the other case, the use of linkage politics was disruptive, showing how easily relationships of complex interdependence can be affected.

Researchers will find other, perhaps less dramatic examples of how a key individual or group redefined the space for actors to work in and move through the welter of jurisdictions with authority over the borders to achieve a policy goal. More subtle factors, such as flagging energies to sustain and deepen long-term commitments—the current attitude in Washington that "NAFTA has been done" being a case in point—or the corrosive effects of budget-cutting on key staff, or even the loss of momentum after the easier targets have been met can affect outcomes. Different levels of wealth and capacity are complicating factors. Given different political systems, the roadmap is often far from clear, but need finds opportunity or, in Tony Hodge's felicitous phrase, "where there's a will, there's a way."

NAFTA provides the large umbrella under which partnerships can flourish, but it is still too soon to know whether they will for certain. And as local and regional groups continue to find ways to input their needs, one asks: Will the inclusionary processes being developed on the borders facilitate the emergence of a North American community?

LESSONS AND RECOMMENDATIONS

Conflict is inevitable; it serves as an engine of change. Cooperation is possible and desirable, given agreement on basic goals and specific targets for attainment. But cooperation also is essential if the public policy goal of shared environmental management is to progress among three nations of such different size and power.

"The environment bears no passport" is a truism amply borne out in the case studies. In the examples of the Great Lakes and Paso del Norte, institutional and cultural differences did not preclude effective cooperation among different jurisdictions. Given the will, officials, citizens, and researchers found ways to implement their shared agenda. Missing overall are coordinating mechanisms to nudge, shape, and prioritize research and governmental programs already in existence. Such coordination well may turn out to be the most important function of the CEC, one of the first and most innovative continental institutions.

The social base for a North American community is being created on the borders, a highly significant development because the new institutions cannot work and evolve without a sense of community. What is unclear is whether governments in this age of austerity have the will to move beyond reactive policies to implement, fund, and sustain the more difficult goals of ecosystem restoration and sustainability. In the words of Hodge and West, "Will the progress of the past twenty-five years be replaced by institutional apathy and a falling away from the ambitious, even noble, goals of the ecosystem management approach that was enshrined in the 1978 Great Lakes Water Quality Agreement?" (see chapter three).

As transboundary environmental issues proliferate and grow more complex, there will be less recourse to federal authorities, who are prone to bureaucratic delays and political inertia, and more direct engagement by officials at the subfederal level, who are closer to the problems. The British Columbia–Washington Environmental Cooperation Council is a case in point. Often the best solutions are those found at the local level, where stakeholders have a direct interest in cooperative approaches and shared solutions. The looming crises over water are a testing ground. Nonetheless, it bears repeating that what is needed is not less government, but more coordinated government at all levels, in tandem with more public participation.

While initiatives at the state and local levels are important, the federal role is still vital. After all, the CEC is fully dependent on agreement among the three federal states, and the same can be said for the International Joint Commission on the U.S.-Canada Border, the International Boundary and Water Commission, the Border Environmental Cooperation Commission, and the North American Development Bank. A regional issue such as water can be addressed only at the federal level, in cooperation with other groups and jurisdictions. Much can be done by municipal officials to conserve water and prevent pollution, but the federal role is critical. For example, at present there is no treaty which addresses the allocation and management of groundwater, a crisis on the southern border. The 1944 water treaty will have to accommodate the explosive growth of Nevada, Arizona, California, and Baja California Norte. Here, too, a water crisis is looming, and government cannot solve it alone.

Business interests on both sides of the Rio Grande will need to get involved. How will El Paso and Ciudad Juárez resolve their water crisis other than through a public-private partnership? Environmental problems wax with the growth of people and maquilas, but the response from business so far has been too little, too late. Moreover, much needs to be done to promote environmental education at the firm and plant levels. Promoting such concepts as eco-efficiency and sustainable business practices has begun, the ISO 14000 training sessions being a case in point.

If business and industry are to avoid environmental conflicts, it is absolutely necessary that they become open to greater transparency in matters of environmental impacts and potential damage. In Canada and the U.S., this is fostered by community-right-to-know legislation, an area in which Mexico still lags, although recent changes in the Environmental Law well may promote greater public access to filed EISs.[34] The multistakeholder consultation process set up recently to deal with the proposed saltworks at San Ignacio also is a significant step toward avoiding conflict and building consensus at the early stages of project design. (However, Mexican environmentalists are being wary.) Conflict in Québec was defused by crafting the Great Whale Forum, but only late in the game.

Demands for the siting of hazardous waste facilities—now flat or declining in Canada and the U.S.—are likely to grow with North America's economy. Recognizing the growing power of NIMBY, one solution local governments have devised is to put the management of hazardous waste treatment facilities under control of the public sector, where it is subject to greater public scrutiny and accountability.

Such is the case of the Gulf Coast Waste Disposal Authority in Houston, a quasi-governmental, independent, nonprofit corporation which functions as a public utility with powers to issue tax-exempt bonds and to condemn land needed to site its facilities.

In Alberta, issues relating to the siting of the Swan Hill incinerator were addressed by involving local and regional interests from the start, so that stakeholders perceived that they were not in a game of clear winners and losers. Through the process of consultation, they came to realize that they all could reap benefits from cooperation.[35]

To overcome NIMBY deadlock, the Southern California Waste Management Authority has promoted public participation in siting decisions by the citizenry and by health and environmental NGOs, reserving the right to make decisions for them if local participants cannot agree among themselves. Although this approach—called YIMBY, or "Yes, in my backyard"—still is largely unproven, it has the potential to replace NIMBY-based parochialism with the ethic of risk sharing and equity in resolving facility-siting disputes across North America. This can be accomplished through greater public-private cooperation.

If any lesson emerges loud and clear from the case studies, it is that the range of interests, groups, and institutions with a stake in environmental management is widening. Local people demand recognition, be they Cree Indians faced with being flooding out of their traditional lands by a Québec power dam, or Ciudad Juárez brickmakers struggling to make a living on very small margins, or Americans of Hispanic descent defending their aquifers against hazardous waste landfills. Having recourse to transnational ENGO networks, their impact is magnified by the inside/outside dynamic that is the hallmark of this new environmental activism.

Many community-led environmental action groups begin with homeowners banding together to protest a local hazard, then they expand their horizons. A case in point is the Grupo de los Cien, which became active protesting the expansion of the Mexico City airport on Lake Texcoco, then expanded its purview to such issues as the Laguna Verde nuclear research station, the Paradise Reef pier, the Silva Reservoir, and protection of the Lacondan rainforest in Chiapas. As the significance of formal boundaries among local, bilateral, and regional-transnational jurisdictions erodes, these groups have more institutional space in which to maneuver and to grow.

In the current political climate, which favors dealing with environmental issues at subnational levels and allowing more operational flexibility for private businesses, it is clear that ENGOs, governments, and business have much to gain by working together in public-private partnerships. If business is challenged to promote environmental performance based on best practices and exceeding mere compliance, ENGOs are challenged to go beyond raising issues to become part of the solution. This role can be uncomfortable, particularly for membership-based organizations dependent upon crisis appeals for funding.[36] And, while they concentrate on protecting the environment, ENGOs must recognize the imperative for developing countries to create wealth and the need for everyone to address the North-South divide. The coming interface between NAFTA and Mercosur will be a test.

Given the environmental challenges facing North America and the public mandate for action, we hope that leaders at all levels of government will respond with vigor. Moreover, a social base is emerging that is conducive to supporting public-private partnerships and a North American community. But will the momentum of environmental management be slowed by funding cutbacks, just when it is on the threshold of new achievements?

If these developments foster a climate that favors the creation of partnerships and cooperation, it is well to remember that, without disputes and conflict, there would be no environmental progress and we would not be in the promising situation we encounter today. Increasingly this "we" is defined trinationally. Hopefully this book, itself a product of trinational cooperation, will add to the pool of shared experience, awareness, and knowledge that we must have to better manage and protect the environment of North America.

NOTES

1. Helen Ingram, Nancy K. Laney, and David M. Gillian, *Divided Waters: Bridging the U.S.-Mexican Border* (Tucson: Univ. of Arizona Press, 1995), 4–5.
2. *New York Times*/CBS polling data, cited in Norman J. Vig and Michael J. Basso, eds., *Environmental Policy in the 1990s*, 2d ed. (Washington, D.C.: Congressional Quarterly Press, 1994), 32. *Reforma* poll, cited in "Una problema de todos," *La Reforma* (Mexico City), Mar. 17, 1995, 3B. Mexican and American opinion on a wide variety of issues is tracked in the *Los Angeles Times/Reforma* National Surveys, most recently in Study No. 380/381, for August, 1996. Canadian polling data show strong support for the environment: Keith McArthur, "Canadians Put Environment First," *Globe and Mail*, June 23, 1997, A4.
3. *Política ambiental en México: El papel de las organizaciones no gubernamentales* (Mexico City: Fundación Friedrich Ebert, 1991), table 20, p. 139.

4. David Brower, comments delivered at NAMI workshop, in *The Role of Environmental Non-Governmental Organizations (ENGOs) in North America,* Santa Fe, N.Mex., Feb., 1995. Brower is founder of Friends of the Earth and dean of U.S. environmentalists.
5. On Oct. 15, 1996, after a year and a half of public consultations and internal politicking by SEMARNAP, the Mexican Attorney General's Office for Environmental Protection, and the environmental committees of the Mexican Chamber of Deputies and Senate, President Zedillo signed into law the long-awaited revisions to the 1988 General Law for Ecological Equilibrium and Protection of the Environment. Mexican ENGOs and business groups are each, from their own positions, treating the changes warily. See the excellent analysis of those changes in John Garrison, "Environmental Law Revised, but Will Greater Environmental Protection Follow?" (typescript), Oct., 1996. Garrison is an environmental consultant working for the Mexican Center for Environmental Law (CEMDA), a Mexican ENGO. See also Gerardo Ramón, "Preven penas de prisión ante delitos ecológicos," *Reforma,* Oct. 11, 1996, p. 2. But, as Betty Aridjis points out, "The law also weakens environmental protection, through virtual elimination of the requirement for full EISs by most new industry, and by such changes as the one which now allows development in biosphere reserve buffer zones as long as it 'benefits the local population.'" Betty Aridjis E-mail to John Wirth, Jan. 7, 1997.
6. Garrison, "Environmental Law Revised"; United Press International (UPI), "Mexico Environmentalists Win Appeal," Nov. 19, 1996. The federal court ruled against SEMARNAP in a suit brought by the Grupo de los Cien concerning weakening of the EIS requirement.
7. Proceedings of a conference on "Renewing Federalism in North America: Diversity of Peoples, Community of Purpose," cosponsored by North American Institute and North America Forum, at Stanford Univ., Stanford, Calif., Mar. 22–24, 1996; available in print under the same title from NAMI and on the Internet at www.santafe.edu/~naminet/fedcon.html.
8. Gerry Mikol, New York Great Lakes Program coordinator, and Jim Haney, a spokesman for the Wisconsin attorney general, quoted in David Poulson [Newhouse News Service], "Great Lakes Water War Renewed," *Mexico City News,* July 25, 1995.
9. Lynton Keith Caldwell, "Conjectures on Continental Cooperation: Transboundary Geopolitics in North America," paper for the North American Institute's "Forum on Borders and Water: North American Waters Issues," Santa Fe, N.Mex., June 26–27, 1992, p. 5.
10. In a precedent-setting move, the three countries agreed to have the CEC investigate whether Mexican laws were complied with in the permitting process for Cozumel Pier. Under Article 14 of the North American Agreement for Environmental Cooperation, the request was brought to the CEC by three Mexican ENGOs: CEMDA, the Grupo de los Cien, and a local group, the Natural Resources Protection Committee. Activists from Greenpeace, the Green Party of Mexico, and CEMDA closed the construction site for two days in Feb., 1996. CEC, *Final Factual Record of the*

Cruise Ship Pier Project in Cozumel . . . (Montréal: CEC, Oct., 1997), is available in print and on the Internet. In fact, ENGOs brought both the Paradise Reef and the Silva Reservoir cases to the CEC; in both instances, the Group de los Cien and CEMDA initiated litigation, with Audubon joining them on Silva and a local group joining the Cozumel suit.

11. Lowry McAllen, "The New Environmentalists," in *Business Mexico* (Oct., 1996): 34–35. Betty Aridjis gives a different view: "The government did hold a public hearing about the saltworks, chiefly as a show for the scientific advisory committee. The process was seriously flawed by the last-minute nature of the hearings and the participation stacked in favor of the project. I don't think the decision to authorize construction of the new saltworks will be taken on the basis of a 'consensus'; SEMARNAP has made it very clear that it will make its own decision, regardless of what the scientific committee, or anyone else, thinks of the new EIS," Aridjis E-mail to Wirth. The Grupo de Los Cien is working with NRDC to mount a new campaign.

12. William H. Bolin, "The Transformation of South America's Borderlands," in *Changing Boundaries of the Americas* (La Jolla, Calif.: Center for U.S.-Mexican Studies, Univ. of California at San Diego, 1992), 169.

13. Victor M. Toledo, "Latinoamérica: Crisis de civilización y ecología política," *Gazeta Ecológica* 38 (Spring, 1996): 13–22.

14. "Lake Sagaris, Chile: Environment vs. Business?" *Latin Trade* (June, 1997): 14.

15. David Pillin, "Menem Embarrassed by Anti-Dam Vote," *Financial Times* (Oct. 11, 1996), 7. The original Hidrovía project is profiled in Laurie Goering, "South American Canal: Financial Boon or Wetlands Danger?" *Chicago Tribune* (Oct. 15, 1995), 7. Redesigned to require less dredging and enhanced by better river controls administered by an international commission, the new, scaled-down project was presented in late 1996. Landlocked Bolivia is the big winner, and Buenos Aires will see more port traffic; but Brazil, with good highway connections to Atlantic ports, has much less need for Hidrovía. All things considered, the project may not be cost-effective. Paulo Prado, of Conservation International, phone conversation with John Wirth, Oct. 29, 1996. See also Environmental Defense Fund and Fundação CEBRAC, *The Hidrovía Paraguay-Paraná Navigation Project; Report of an Independent Review* (Washington and Brasília, July, 1997).

16. Jose Roberto Borges, Rainforest Action Network, San Francisco, conversations with John Wirth, Nov., 1996.

17. Rodolfo Cardona, "Falsos ecologistas tratan de manipular la opinión pública," *Universal* (Caracas), June 18, 1997, 1, 19. Carlos Camacho, "Fishermen Will Block Channel," *Caracas Daily Journal,* Mar. 30, 1997, 1.

18. "Colombia y Venezuela buscan salidas ambientales para la Sierra de Perija," *El Nacional (Caracas),* Oct. 27, 1996, A36.

19. "Ecuador to Pursue Environmental Suit Against Texaco," *Mexico City Times,* Nov. 29, 1996, 4. On Nov. 12, Judge Jed Rakoff in the U.S. District Court in New York rejected the lawsuit by Indians and settlers in the

Amazon region who claimed Texaco caused millions of dollars worth of damage in the rainforest by improper containment of hazardous waste. On the smelter, see Calvin Sims, "In Peru, a Fight for Fresh Air, Residents Say American Copper Plant Sickens Them," *New York Times*, Dec. 12, 1995, D1; and Comité de Defensa de la Salud de Ilo, "Daños causados por la Southern Peru Copper Corporation a la población [by sulphur dioxide pollution]," press release, Lima, Peru, Sept. 3, 1996. After being turned down by Judge Janis Graham Jack in U.S. District Court in Corpus Christi, the plaintiffs sought to gain standing in a Louisiana court.

20. Samper is current president of the Non-Aligned Movement, made up of 101 nations. "Latin Nations Oppose Linking Human Rights to Trade," *Mexico City Times*, Oct. 25, 1996, 11.

21. Colombia and Venezuela already have a free trade agreement with Mexico, and the three countries are reported to be making environmental protection a top priority. Among the major trading nations of South America, Colombia and Venezuela, after Chile, may be the first to accept the trade and environment linkage; *Caracas Daily Journal*, Nov. 16, 1996, 1–2.

22. Chile signed the bilateral trade agreement with Canada in Nov., 1996. John D. Wirth, "The Trade-Environment Nexus," revised draft of a paper presented at the "Third Forum NAFTA/Mercosur," organized by the International Relations Program, Univ. of São Paulo, Brazil, Sept. 4, 1995.

23. "Only if all parts of a society (industry, trade unions, ENGOs, and the public at large) participate more actively in environmental decisions—and the environmental plans do much to promote such participation—will it be possible to make goals more consistent. Today, nearly all OECD countries (except Mexico) encourage public participation and disseminate information on the state of the environment, and some even recognize the right of environmental information to all." Michel Potier, "Integrating Environment and Economy," *OECD Observer*, no. 198 (Feb.–Mar., 1996): 6. The new environmental law passed by Mexico's Congress addresses this issue.

24. McAllen, "New Environmentalists," 32, 34.

25. Grace Wever, *Strategic Environmental Management: Using TQEM and ISO 14000 for Competitive Advantage* (New York: John Wiley and Sons, 1996), 21.

26. Richard Kiy, "The Gulf of Mexico: North America's Maritime Frontier," in *Identities in North America: The Search for Community*, ed. Robert L. Earle and John D. Wirth (Stanford, Calif.: Stanford Univ. Press, 1995), 175–76.

27. The agencies involved are the Texas General Land Office, which is promoting the Transborder Information Program (a GIS inventory), and the TNRCC, which is giving technical assistance for pollution prevention to border area industries. Such initiatives complement efforts by such ENGOs as EDF and the Texas Center for Policy Studies.

28. Charles Colgan and Jaime Plumstead, "Economic Prospects for the Gulf of Maine Region," White Paper, Gulf of Maine Council on the Marine

Environment, Oct., 1995; available on the council's website, http://rosby.unh.edu/edims/documents/economic/economic.html.

29. Rick Careless, "Tatshenshini, Protecting North America's Wildest River," in *To Save the Wild Earth; Field Notes from the Environmental Frontline* (Vancouver: Raincoast Books, 1997), 167–99.

30. "Study Condemns Arctic Oil Drilling," *Washington Post*, Aug. 27, 1995; Raymond D. Cameron, *Distribution and Productivity of the Central Arctic Caribou Herd in Relation to Petroleum Development: Case History Studies with a Nutritional Perspective* (Anchorage: Alaska Department of Fish and Game, Dec., 1994).

31. U.S.-Mexico Border Governors Conference, pre-briefing notes, April, 1992, mimeographed. Mary May Simon, *Inuit: One Future—One Arctic* (Peterborough, Ont., Canada: Cider Press, 1996), chs. 1 and 2, quotation on 26–27.

32. Tony Homer Allen-Mills, "Louisiana Blacks Win Nuclear War," *Sunday Times of London*, May 11, 1997, p. 19.

33. Red River Flood Links, www.brandonu.ca/~education/flood/Welcome.html.

34. Garrison, "Environmental Law Revised," relates: "Citizen participation and access to information is not limited to the audit and EIA processes. Under the new law SEMARNAP is charged with creating a National Environmental and Natural Resource Information System which will contain inventories of natural resources, atmospheric emissions, water discharges, and hazardous materials and wastes based on environmental permits, licenses and authorization. Citizens are also granted the right to request in writing 'environmental information' from Government authorities."

35. Daniel Mazmanian and David Morell, "The NIMBY Syndrome: Facility Siting," in *Environmental Policy in the 1990s*, 2d ed., ed. Norman Vig and Michael Kraft (Washington, D.C.: Congressional Quarterly Press, 1994), 245.

36. These issues were discussed at a symposium on "The Role of Environmental Non-Governmental Organizations (ENGOs) in North America," sponsored by the North American Institute, Santa Fe, N.Mex., Feb., 1995. See also n. 31 to the introduction to this volume.

NORTH AMERICAN ENVIRONMENTAL CHRONOLOGY

RICHARD KIY AND JOHN D. WIRTH,
WITH THE ASSISTANCE OF BETTY ARIDJIS,
JOHN GARRISON, AND JANINE FERRETTI

1885	New York State, in cooperation with the Province of Ontario, creates the Niagara Reservation, protecting Niagara Falls.
1889, Mar. 1	U.S.-Mexico Convention signed to deal with changes in the beds of the Rio Grande and Colorado River.
1891	National Irrigation Congress organized in the U.S.
1892	Sierra Club founded by 162 Californians, with John Muir as president.
1894	First International Irrigation Congress held in Denver, Colorado, with representatives of both Canada and Mexico. Here the notion of a North American Conservation Commission first is put forward.
1898	Gifford Pinchot named head of the U.S. Division of Forestry.
1899	River and Harbour Act establishes the first legal basis for banning pollution of navigable waters in Canada.
1905	National Audubon Society formed in U.S.
1905	U.S.-Canada International Waterways Commission established to advise the two countries on water and boundary issues.
1906, May 21	U.S.-Mexico Convention Providing for the Equitable Distribution of the Waters of the Rio Grande for Irrigation Purposes.
1909, Jan. 11	Treaty Between the U.S. and Great Britain Relating to Boundary Waters. Establishes the International Joint Commission on the U.S.-Canada Border (IJC).
1909, Feb.	North American Conservation Conference held in Washington, D.C.
1909	Founding of the Canadian Commission on Conservation. A research clearinghouse for the natural resources work done by federal and provincial governments, it was disbanded in 1921.
1910	Ballinger-Pinchot controversy disrupts the conservation movement.

1913	U.S. Migratory Bird Act, designed to regulate hunting, generates controversy. Spring hunting and marketing of hunted birds prohibited (1918 treaty with Canada will solidify regulations). Act also prohibits importing into the U.S. wild bird feathers for women's fashions.
1916, Dec. 7	U.S.-Canada Convention for the Protection of Migratory Birds.
1918	U.S. Migratory Bird Treaty Act implements the 1916 treaty with Canada to restrict hunting of migratory species.
1924	Oil Pollution Control Act in U.S.
1925, July 17	Treaty in Regard to the Boundary Between the U.S. and Canada.
1927–41	Famous Trail Smelter (lead and zinc) case, in which U.S. complained about transborder pollution from British Columbia to Washington State. Case led to arbitration and the first international ruling that "the polluter pays."
1935	Wilderness Society formed by Aldo Leopold and Robert Marshall in U.S.
1936	National Wildlife Federation founded in U.S. In the 1980s, it would have 4.6 million members.
1936, Feb. 7	U.S.-Mexico Convention for the Protection of Migratory Bird and Game Mammals.
1936	U.S.-Mexico-Canada Migratory Bird Treaty.
1940	Creation of the U.S. Fish and Wildlife Service consolidates federal protection and propagation activities.
1944, Nov. 14	U.S.-Mexico Treaty Relating to the Utilization of Waters of the Colorado and Tijuana Rivers and the Rio Grande. Establishes International Boundary and Water Commission.
1946	U.S. Bureau of Land Management established to consolidate administration of the public domain.
1946, Dec. 2	International Convention for the Regulation of Whaling.
1947	Defenders of Wildlife founded in U.S.
1948	Federal Water Pollution Control Law enacted to regulate waste disposal in U.S.
1949	Aldo Leopold's *A Sand County Almanac* is published posthumously.
1950	Poza Rica killer smog incident leaves twenty-two dead, hundreds hospitalized in Mexico.
1950s	Great Lakes fisheries collapse due to overfishing, industrialization, and environmental mismanagement.
1952, June 30	St. Lawrence Seaway Agreement in Canada-U.S.
1953, Nov. 12	U.S.-Canada Agreement Regarding the Establishment of the St. Lawrence River Joint Board of Engineers.
1954, Sept. 10	U.S.-Canada Convention on Great Lakes Fisheries.
1960, Nov. 15	United Nations Food and Agriculture Organization (FAO) Resolution Establishing the North American Forestry Commission.
1961	World Wildlife Fund founded.
1961, Jan. 17	Treaty between the U.S. and Canada on cooperative development of water resources of the Colombia River Basin.

1962	Rachel Carson's *Silent Spring* published.
1964, Jan. 22	U.S.-Canada Agreement establishing Roosevelt Campobello International Park.
1967	Santa Barbara oil drilling platform oil spill off coast of California sparks major public concern about environmental protection. It followed the stunning Torrey Canyon supertanker oil spill off the coast of England just months before. Oil spill response teams, oil spill legislation, oil spill cleanup funds, and oil tanker inspection and maintenance requirements will ensue.
1967, Apr. 5	U.S.-Canada Convention on Great Lakes Fisheries.
1967	Environmental Defense Fund established to fight for conservation issues through the legal system and science.
1967	Environmental Defense Fund, Sierra Club, and David Anderson launch first court action requiring environmental impact assessment in U.S. It dealt with the Alaska Pipeline and Atlantic Richfield in Alaska.
1968–69	Canada's largest environmental organizations (as opposed to conservation organizations, which were created earlier) formed. Among them: Pollution Probe, SPEC British Columbia, STOP Québec, STOP Alberta, Recycling Councils (one for each of the major provinces).
1969–71	Canada creates its key environmental law associations, including Canadian Environmental Law Association, Canadian Environmental Research Foundation (now called Canadian Institute for Environmental Law and Policy), West Coast Environmental Law Association, and Alberta Environmental Law Association.
1969	Pollution Probe and SPEC created in Canada.
1969	Friends of the Earth founded in U.S. by David Brower after his ouster from the Sierra Club.
1969	"Don't Make a Wave Committee" created for Amchitka protest by Canadian environmentalists. It becomes Greenpeace International.
1970s	Abilone Alliance organized in California to oppose construction of nuclear power plants.
1970, Jan. 1	U.S. National Environmental Policy Act (NEPA) signed.
1970, Apr. 22	Earth Day held in U.S. Environmental Action formed to coordinate Earth Day, lobby aggressively, and alert voters to the "dirty dozen" companies with the worst pollution records.
1970	U.S. League of Conservation Voters established by leaders of national environmental groups to track voting records and policy decisions of members of Congress and the executive branch.
1970	Canada, with future Environment Minister John Fraser, prevents the flooding of the Skagit River in British Columbia by a new dam in Washington State, by referring the matter to IJC.
1970, Nov. 20	Treaty to resolve pending boundary differences and maintain the Rio Grande and Colorado River as the international boundary between the U.S. and Mexico.

1970	National Resources Defense Council established in U.S.
1970, Dec. 2	U.S. Environmental Protection Agency (EPA) created by President Richard Nixon.
1970, Dec. 31	U.S. Clean Air Act amendments expand scope of the 1967 act.
1971	Love Canal near Niagara River, New York State, revealed to be the site of buried chemical wastes, endangering the health of local residents near Canadian border.
1971	Southwest Research and Information Center created in Albuquerque, New Mexico, to link the knowledge of environmental experts with community groups to support and promote local environmental issues.
1971	Greenpeace incorporated in Vancouver, B.C.
1971	Broad use of DDT and related chlorinated organic pesticides banned in U.S. and Canada.
1971	Department of Environment Act of 1971, Canada, combines federal entities responsible for various aspects of the environment, including resource management and regulatory functions, into newly formed department called Environment Canada.
1972	Environmental Policy Center founded in U.S., later called Environmental Policy Institute.
1972, Apr. 15	Great Lakes Water Quality Agreement in Canada-U.S.
1972	U.S. Clean Water Act.
1972	U.S. Ocean Dumping Act.
1972	U.S. Coastal Zone Management Act empowers states to lead in planning and regulation.
1972	United Nations Conference on the Human Environment held in Stockholm, Sweden.
1973	U.S. Endangered Species Act.
1973	Major freeways proposed to bisect Toronto (the Spadina) and Vancouver (the Third Crossing) prevented after a three-year battle, forcing the municipalities into stronger rapid and public transit programs.
1973	Canada-U.S. Environmental Council (CUSEC) created. It operated for seven years to resolve differences among Canadian and U.S. environmental groups who were beginning to say, "Not in my backyard; put it in your backyard." The key issue was the location of the Alaska supertanker oil terminal: Long Beach, California; Cherry Point, Washington State; or Kitimat, British Columbia? Canadian groups wanted it located in Long Beach, not Cherry Point or Kitimat. U.S. groups wanted it in Canada. The Thompson Inquiry stopped this project in 1976.

CUSEC was made up of the largest environmental and conservation organizations from both countries, including National Resources Defense Council, Sierra Club, Audubon Society, Friends of the Earth, and the U.S. National Wildlife Federation. Additional issues included acid rain, Skagit River damming, nuclear power plant locations near borders, and the Garrison

	Diversion (one of the U.S. Army Corps of Engineers' largest proposed water damming and irrigation schemes, it was to be located in North Dakota and would have flooded a large section of lower Manitoba).
1973, Aug. 30	U.S.-Mexico International Boundary and Water Commission, Minute 242, provides definitive solution to international problem of the salinity of Colorado River.
1974	U.S. Safe Drinking Water Act.
1974, June 19	U.S.-Canada agreement providing for joint pollution contingency plans for spills and other noxious substances.
1975	Justice Thomas Berger completes Canada's first full environmental assessment. It dealt with proposed construction of a 2,000-mile-long natural gas pipeline proposed by a consortium of twenty-six U.S. and Canadian energy companies to take natural gas from the Canadian Arctic's Beaufort Sea to the South and Midwest of the U.S. The assessment resulted in refusal to grant approval for the undertaking. In Canada this four-year process set off a wave of environmental assessment law and funding for intervenors. Berger provided the first intervenor funding in the history of Canadian environmental hearings.
1976	Resource Conservation and Recovery Act in U.S.
1976	Ontario becomes first jurisdiction in Canada to pass an environmental assessment act.
1976, Oct. 12	U.S.-Mexico-Canada Agreement for the Protection of Plants.
1976, Nov. 24	U.S.-Mexico Agreement Concerning Certain Maritime Boundaries.
1976	United Nations Conference on Human Settlements held in Vancouver.
1976	U.S. Toxic Substance Control Act.
1977	Inuit Circumpolar Conference established in Barrow, Alaska.
1977	Use of Agent Orange herbicide 2,4,5-T banned in U.S. and Canada.
1978, May 4	U.S.-Mexico Treaty on Maritime Delimitation.
1978, Sept. 21	Organization for Economic Cooperation and Development (OECD) Council recommendation on strengthening international cooperation on environmental protection in frontier regions.
1978, Nov. 18	Revised Great Lakes Water Quality Agreement of 1978.
1979, Mar. 28	Three-Mile Island nuclear reactor failure.
1979	First recorded event of environmental activism in Mexico. Comité de Defensa Ecológica de Michoacán organizes protests against proposal to build a nuclear plant on the shores of Lake Pátzcuaro, State of Michoacán. Lobbying efforts eventually will kill the planned project.
1980s	Earth Island Institute founded in U.S. by David Brower.
1980s	Wise-Use Movement founded in U.S. to advocate using the environment to support human needs and deregulating government legislation that restricts industry and exploitation of natural resources.

1980, Aug. 5	U.S.-Canada Memorandum of Intent Concerning Transboundary Air Pollution.
1980	Coalition on Acid Rain formed by Canada's John Fraser and others to lobby the U.S. government as well as to lobby in Canada.
1980	U.S. Comprehensive Environmental Response, Compensation, and Liability Act (Superfund).
1981	Canadian nongovernmental organizations first granted *amicus curiae* status in a U.S. federal court, in a case dealing with the Hyde Park landfill, *Pollution Probe v. Hooker Chemical, Occidental Petroleum*. Pollution Probe is represented by the Canadian Environmental Law Association.
1981	Anne Gorsuch resigns as head of the U.S. EPA after revelations of mismanagement.
1982	Mexico's Secretariat for Ecology and Urban Development and Housing (SEDUE) established.
1982	Great Lakes United created, a formal binational citizen coalition on the U.S.-Canada border.
1982	Organización Ribereña Contra la Contaminación del Lago de Pátzcuaro (ORCA) founded by fisherman and peasants.
1983	James Watt resigns as secretary of U.S. Department of the Interior under mounting public criticism for antienvironmental policies.
1983	Centro de Estudios Sociales y Ecológicos established to support research work for ORCA in the Lake Pátzcuaro region of Mexico.
1983 Aug. 14	U.S.-Mexico Agreement on Cooperation for the Protection and Improvement of the Environment in the Border Region, otherwise known as the La Paz Agreement, signed by Presidents Reagan and de la Madrid.
1984, Apr. 2	U.S.-Canada Treaty Relating to the Skagit River and Ross Lake and the Seven-Mile Reservoir on the Pend d'Oreille River.
1984	Red Alternativa de Eco-Contaminación established in Mexico.
1984, Oct. 12	U.S.-Canada Delimitation on the Maritime Boundary in the Gulf of Maine Area.
1984, Sept.	PEMEX storage tanks at San Juan Ixhuátepec, in the State of Mexico, explode, killing thousands of people.
1984, Dec. 5	U.S.-Mexico Agreement Establishing the Joint Committee for Conservation of Wildlife.
1985	Canadian Acid Rain Reduction Targets established; goal is, by 1994, 50 percent down from 1980 allowable limits.
1985	Major discovery of toxic chemical blob from Dow Chemical Canada's facility at Sarnia on the St. Clair River. Contributes to speedy creation of the Responsible Care program in Canada in 1987. This latter will be adopted by the U.S. and the international chemical industry. Will result in creation of IJC Remediation Action Plans (RAPs) for cleaning up toxic hot spots in the Great Lakes Basin. Will result in the first modern set of comprehensive water pollution control laws.
1985	Unión Popular de Ixhuátepec, A.C., founded to focus attention

	on deaths of Ixhuatepec residents killed in the PEMEX explosion and lack of health and safety precautions at plant.
1985	Ontario uses $25 million voluntary contribution from private sector to roll out North America's first very large curbside residential/commercial "Blue Box Recycling" program, which today serves 3 million homes in Ontario. This model will be copied throughout the U.S. and Europe.
1985, Mar. 1	Grupo de los Cien Internacional (Group of 100) formed by Homero Aridjis and others to protest pollution and environmental negligence in the Valley of Mexico and the lack of official and private-sector attention to these problems. The first concrete victory will occur in June, when the group successfully opposes expansion of the Mexico City Airport into Lake Texcoco.
1985, June 24–28	Universidad Nacional Autónoma de México hosts first national conference on social activism and the environment.
1985	Mexico City earthquake focuses attention on lack of environmental and safety standards, as thousands die in the quake.
1985, Oct. 17	U.S.-Canada Memorandum of Understanding Concerning the Research and Development Cooperation in Science and Technology for Pollution Measurement and Control.
1985, Oct. 17	U.S.-Canada Memorandum of Understanding Regarding Accidental and Unauthorized Discharges of Pollutants Along the Inland Boundary.
1985, Nov.	First National Congress of Mexican Environmentalists.
1985	Spills Bill passed.
1986	U.S. Superfund reauthorized.
1986	U.S.-Canada Agreement Establishing the North American Waterfowl Management Plan (joined by Mexico in 1994).
1986, June 5	OECD Council Decision and Recommendations on Export of Hazardous Wastes from OECD Areas (known as the Basel Convention).
1986	Widespread protests by Mexican environmental community opposing Laguna Verde Nuclear Plant. Protests started late (following Chernobyl disaster), at advanced stage of project, so initiative proves unsuccessful.
1986	Ontario passes Countdown Acid Rain regulations mandating 60-percent cuts in sulphur dioxide emissions by 1994 from four major sources, including INCO Nickel Smelter in Sudbury, the largest single source of sulphur dioxide emissions in North America, and Ontario Hydro.
1986	Pacto de Grupos Ecologistas (Pact Among Environmental Groups) established in Mexico.
1986	Kativik Environmental Quality Commission, created by the James Bay and Northern Québec Agreement to carry out environmental impact assessment in northern Québec.
1986–87, Winter	Worst pollution in Mexico City's history reported. Prompts attention to city's worsening air-pollution problem.

1987	Canadian Transportation of Dangerous Goods Act.
1987	Montréal Protocol on chlorofluorocarbon reduction and phase-out created and headquartered in Montréal.
1987	Canadian National Packaging Protocol adopted by the federal government and the provinces, to achieve a 50-percent reduction of waste by the year 2000.
1987	Mexican Green Party established (later known as the Ecologist Party of Mexico, or PEM).
1988–90	Canada Environmental Protection Act strengthens federal jurisdiction.
1988	Creation of the National Roundtable on the Environment and the Economy in Canada, as well as a number of other provincial roundtables on the environment and the economy, after tabling of the Brundtland Commission Report. The roundtables were recommended by the National Task Force on the Environment and Economy in its 1987 report.
1988, Jan. 11	U.S.-Canada Agreement on Arctic Cooperation.
1988, Mar. 16	U.S.-Canada-Mexico Memorandum of Understanding on exchange of information and cooperation on wetlands and migratory bird refuges; establishment of a Tripartite Committee to develop a strategy for conservation of migratory birds and their habitats.
1988, June 7	Coordinadora Nacional del Movimiento Anti-Nuclear, Mexico, organizes campaign against PRI presidential candidate Salinas because of his failure to oppose Laguna Verde. First public environmentalist attack on a Mexican presidential candidate.
1988	Mexican Generalized Law for Environmental Equilibrium enacted.
1989	Exxon Valdes Oil Spill in Prince William Sound, Alaska, near Canadian waters.
1989, Oct. 3	U.S.-Mexico agreement on cooperation for the protection and improvement of the environment in the metropolitan area of Mexico City.
1989	Ontario is first jurisdiction in Canada to pass legislation phasing out chlorofluorocarbons and ozone-depleting substances (ODP).
1989	Environmental Consortium for Minority Outreach established in U.S. by leading environmental groups to attract and encourage participation by racial and ethnic minorities in environmental organizations.
1989	Due to pressures by environmental nongovernmental organizations, U.S. Congress reapplies an embargo on Mexican tuna first established in 1980.
1989	Mulroney's Conservative Party wins elections in Canada, and that summer the government begins drafting its first comprehensive national environmental plan.
Early 1990s	Save the Georgia Strait Alliance formed in Canada.
1990, Mar. 16	U.S.-Mexico Agreement on Environmental Cooperation.
1990	U.S. community of nongovernmental organizations successfully

	pushes concept of "Dolphin Safe" tuna, prolonging the Mexican tuna embargo.
1990, Sept. 26	President Bush requests U.S. Congressional permission to negotiate a free trade agreement with Mexico.
1990, Nov. 27	U.S.-Mexico Presidential Summit in Monterrey, Mexico, whereby an agreement, in principle, is reached to establish a U.S.-Mexico Free Trade Agreement and to initiate a binational environmental border plan to address common problems along the U.S.-Mexico border.
1990	Canada launches the Green Plan, a comprehensive program to regulate toxics, reduce generation of waste by 50 percent by the year 2000, extend environmental assessment criteria to eleven federal agencies, and promote sustainable development in forestry, agriculture, and fisheries.
1991, Feb. 5	Canada joins free trade negotiations with U.S. and Mexico.
1991, Mar. 1	U.S. President Bush requests "fast-track" negotiating authority for NAFTA from U.S. Congress.
1991, Mar. 7	Letter to President Bush from Sen. Lloyd Bentsen and Rep. Dan Rostenkowski, requesting Bush administration to address environmental, health, and safety standards, along with workers' rights, as a condition of "fast-track."
1991, Mar. 13	U.S.-Canada Agreement on Air Quality, known as "Acid Rain Accord."
1991, May 23	"Fast-track" negotiating authority passed by U.S. Congress, provided that environmental, health, and safety standards and workers' rights are included in NAFTA.
1991, June	NAFTA negotiations begun by U.S., Canada, and Mexico.
1991, July 2	Public Citizen, a consumer action group in U.S., requests that the U.S. Trade Representative prepare an environmental impact statement for NAFTA under the National Environmental Policy Act (NEPA).
1991, Aug. 1	Public Citizen files lawsuit against U.S. Trade Representative regarding requirement of environmental impact statement for NAFTA under NEPA.
1992, Feb. 29	EPA releases Integrated Environmental Plan for the U.S.-Mexico Border, First Stage (1992–94).
1992, Mar. 27	New York cancels contract to purchase hydroelectric power from Hydro-Québec.
1992, Apr. 3	Permit for TITISA, hazardous waste incinerator in Tijuana, Mexico, canceled by SEDUE Secretary Patricio Chirinos.
1992, Apr.	State of Coahuila, Mexico, granted party status in hearing before the Texas Natural Resource Conservation Commission, on permitting of Dryden, Texas, waste site.
1992, May 7	British Colombia–Washington Environmental Cooperation Council established.
1992, June	Earth Summit in Rio de Janeiro, Brazil, with Canada's Maurice Strong presiding.

1992	U.S. Sen. Al Gore's book, *Earth in the Balance: Ecology and the Human Spirit,* published.
1992	Grupo de los Cien and Alert Citizens for Environmental Safety form a coalition to oppose the siting of a low-level nuclear waste dump in Sierra Blanca, Texas.
1992	SEDESOL established as Mexico's new environmental and development agency, replacing SEDUE. Luis Donaldo Colosio is first secretary of SEDESOL.
1992, Aug. 12	Conclusion of NAFTA negotiations by U.S., Canada, and Mexico.
1992, Sept. 14	Public Citizen refiles lawsuit requesting requirement of an environmental impact statement for NAFTA.
1992, Sept. 17	First meeting of the North American Environmental Ministers, at Blair House, Washington, D.C. An agreement, in principle, was reached to establish a North American Commission for Environmental Cooperation, which later became the CEC.
1992, Sept. 17	U.S.-Canada-Mexico Memorandum of Understanding on Environmental Education.
1992, Oct. 6	U.S. presidential candidate Bill Clinton calls for side agreements to NAFTA on labor and environment, in campaign speech at Raleigh, N.C.
1993	Green Group formed to support NAFTA and advocate the trade-environment linkage. This coalition is organized by Audubon Society with National Wildlife Federation, World Wildlife Fund, and National Resources Defense Council, as distinct from the Sierra Club and Friends of the Earth, who opposed NAFTA.
1993, May 4	Letter from U.S. environmental leaders to U.S. Trade Representative Mickey Kantor, requesting a stronger environmental side agreement to NAFTA.
1993, June 30	In Public Citizen's case, U.S. Appeals Court rules in favor of U.S. Trade Representative.
1993, July 22	Address by U.S. House Majority Leader Gephardt on NAFTA, side agreements, and border funding.
1993, Sept. 21	North American Agreement on Environmental Cooperation signed by U.S., Canada, and Mexico, establishing the CEC.
1993, Sept. 21	U.S.-Mexico Border Environmental Cooperation Agreement signed, establishing the Border Environmental Cooperation Commission and the North American Development Bank.
1993, Nov. 17	NAFTA ratified by U.S. Congress.
1994, Dec.	SEMARNAP established, consolidating SEDESOL and Mexican fisheries agency (Secretaría de Pesca).
1995, June 14	CEC decides to investigate bird deaths at Silva Reservoir in Guanajuato State, Mexico; the trinational scientific report is a model of its kind.
1996, Feb. 12	Greenpeace ship, *Moby Dick,* blockades Cozumel Pier development, drawing international attention to this environmentally controversial project.

1996, Apr.	Village of Tepoztlán defeats golf course developers. Protests lead to death of two townspeople.
1996, June 7	CEC recommends preparation of a factual record on Mexican compliance with environmental impact statement in permitting of Cozumel Pier, in response to a submission by the Grupo de los Cien, CEMDA, Audubon Society, and a local group.
1996, Aug. 1	Mexico's Environmental Secretary Julia Carabias protests CEC actions in Cozumel Pier case, citing an infringement of Mexican sovereignty, but agrees with her Canadian and U.S. counterparts on preparation of a factual record.
1996, Sept. 19	Arctic Council established by the eight circumpolar nations, including Canada and the U.S. 1996 Chemical explosion at PEMEX petrochemical facility at Cactus, Mexico, leaves six dead, focusing attention on lack of health, safety and environmental controls at the company's facilities. SEMARNAP calls for nationwide environmental audits of PEMEX facilities.
1996, Oct.	U.S.-Mexico Border XXI Framework Document issued, addressing border environment, natural resources, and public health issues.
1996, Oct. 15	Mexican Congress amends the country's basic Environmental Law (Ley General de Equilibrio y Medio Ambiente), providing for greater public access to information, among other reforms.
1996	Binational alliance including Texas Center for Policy Studies and Grupo de los Cien mobilizes to oppose possible extension of the U.S. Intercoastal Waterway with a Mexican counterpart which would bisect the Laguna Madre, a major fish and bird habitat.
1996, Nov. 18–19	Environmental officials from the ten Mexico and U.S. border states meet in Austin, Texas, to promote a regional action plan.
1997	Grupo de los Cien wins federal court suit against SEMARNAP concerning weakening of the environmental impact statement requirement in the new law. For the first time, a Mexican court recognizes the legal right of a group or an individual to challenge a governmental ruling or regulation; before, only parties with a direct interest were granted legal standing.
1997, June	Environmental ministers approve program by CEC to study long-range transport of pollution in North America.

SELECTED BIBLIOGRAPHY

Borland, H. *The History of Wildlife in America*. Washington, D.C.: National Wildlife Federation, 1975.
Cohen, M. *History of the Sierra Club*. San Francisco: Sierra Club Books, 1988.
Demmers, Jolle, and Barbara Hogenboom. "Popular Organization and Party Dominance: The Political Role of Environmental Nongovernmental Organizations in Mexico." Master's thesis, Faculty of Political and Sociocultural Sciences, University of Amsterdam, Netherlands, November, 1992.
Doern, G. Bruce, and Thomas Conway. *The Greening of Canada: Federal Institutions and Decisions*. Toronto: University of Toronto Press, 1984. Reprinted, Ottawa: Environment Canada, 1996.
Earle, Robert L. and John D. Wirth. *Identities in North America: The Search for Community*. Stanford, Calif.: Stanford University Press, 1995.
Frazier Nash, Roderick. *American Environmentalism: Readings in Conservation History*. 3d ed. New York: McGraw-Hill, 1976.
Israelson, David. *Silent Earth: The Politics of Our Survival*. Markham, Ontario: Penguin Books, 1990.
Kline, Benjamin. *First Along the River: A Brief History of the U.S. Environmental Movement*. San Francisco: Arcada Books, 1997.
Neuzil, Mark, and Bill Kovarik. *Mass Media and Environmental Conflict: America's Green Crusade*. New York: Sage, 1996.
Sale, K. *The Green Revolution: The American Environmental Movement, 1962–1992*. New York: Hill and Wang, 1993.
Szekely, Alberto. *Establishing a Region for Ecological Cooperation in North America*. Albuquerque: International Transboundary Resource Center, University of New Mexico, Summer, 1992.
Universidad Nacional Autónoma de Mexico. *Movimientos sociales y medio ambiente*. Mexico City: Programa Universitario "Justo Sierra" de la Coordinación de Humanidades, UNAM, June, 1985.

Wall, D. *Green History: A Reader in Environmental Literature, Philosophy and Politics*. London: Routledge, 1994.
Wanners, W. *T. R. and Will: A Friendship That Split the Republican Party*. New York: Harcourt, Brace, 1969.
Weart, S. R. *Nuclear Fear: A History of Images*. Cambridge, Massachusetts: Harvard University Press, 1988.
White, R. "American Environmental History: The Development of a New Historical Field." *Pacific Historical Review* 54 (August, 1995): 297–335.
Worster, Donald, ed. *The Ends of the Earth: Perspectives on Modern Environmental History* (New York: Cambridge University Press, 1988).

CONTRIBUTORS

JAMIE ALLEY is a senior civil servant with the Province of British Colombia, where he was asked to develop the model for the British Columbia-Washington Environmental Cooperation Council. Most recently he served as a special advisor to the Premier's Office on intergovernmental and environmental affairs before turning his attention to fisheries.

CARLOS ANGULO is an attorney with Baker & McKenzie, Cd. Juárez, Chihuahua, Mexico; a business consultant; and a rising star in Mexico's PAN party.

GEOFFREY J. BANNISTER is an economist and assistant professor of economics at the University of New Mexico's Anderson School of Management.

ALLEN BLACKMAN is an economist and a fellow in the Quality of the Environment Division at Resources for the Future, a research institute in Washington, D.C.

DAVID CLICHE is former minister of environment for the Province of Quebec, having been in charge of indigenous affairs for the Quebec provincial government. Previously he organized the multi-stakeholder Great Whale Forum, dealing with hydro-power issues. Currently he is the province's minister of tourism. LUCIE DUMAS is a staff member at Indigenous Affairs.

PETER EMERSON is a senior economist with the Environmental Defense Fund and has been a leader of the multi-stakeholder Ciudad Juárez-El Paso Air Quality Task Force. Previously, he served as vice president for resource planning and economics with the Wilderness Society; principal analyst for the U.S. Congressional Budget Office; and agricultural economist at the U.S. Department of Agriculture.

SERGIO ESTRADA ORIHUELA formerly was director of ecological planning at Mexico's Secretariat of Social Development (SEDESOL) and director of norms and standards for the Secretariat of Ecology and Urban Development (SEDUE). Currently Dr. Estrada is general coordinator of the Linkage Project for Competitiveness and Sustainable Industrial Development, at the Center for Technological Innovation, Universidad Nacional Autonoma de Mexico.

ALFREDO GUTIÉRREZ, M.D., long served as mayor of Del Rio. He spearheaded efforts to defeat three large projects which would have had adverse impacts on Del Rio's aquifer.

R. ANTHONY HODGE is a practicing engineer, teaches Environment and Management at Royal Roads University, and is Adjunct Professor at the University of Victoria. From 1992 to 1996 he served on Canada's National Round Table on the Environment and the Economy. He is co-author of *Great Lakes, Great Legacy?* and *Pathways to Sustainability: Addressing Our Progress*.

MARY KELLY, an attorney based in Austin, is executive director of the Texas Center for Policy Studies and over the past five years has developed and coordinated the highly successful Texas-Northeast Mexico Network on the Environment. Kelly is past chair of the National Advisory Committee on NAFTA Environmental Affairs for the U.S. Environmental Protection Agency.

JOHN KIRTON is an associate professor of political science, research associate of the Center for International Studies, and director of the G7 Research Group at the University of Toronto. He is co-author of *NAFTA's Institutions: The Environmental Potential and Performance of the NAFTA Free Trade Commission and Related Bodies* and co-editor of *Trade and the Environment: Economic, Legal and Policy Perspectives*.

RICHARD KIY currently is director for environment, health, and safety systems at INTESA (an SAIC joint venture information technology services company), Caracas, Venezuela. Formerly he served as a specialist on U.S.-Mexico border affairs with the U.S. Environmental Protection Agency and as acting environmental attache in the U.S. Embassy in Mexico City.

STEPHEN MUMME is professor of political science at Colorado State University where he specializes in comparative and international environmental politics and policy. He is the author of *Apportioning Groundwater along the U.S.-Mexico Border* and coauthor of *Statecraft, Domestic Politics and Foreign Policymaking: the El Chamizal Dispute*. In 1997–98 Mumme was a Fulbright Scholar at the Colegio de Sonora studying state and municipal environmental management in Mexico.

DONALD MUNTON is professor and chair of International Studies at the University of Northern British Columbia. He teaches international environmental policy, Canadian foreign policy, Canadian-American relations, and introductory international studies. He recently published an edited collection, *Hazardous Waste Siting and Democratic Choice*.

ANN C. PIZZORUSSO is director for environmental affairs for Philips Electronics North America Corporation and serves on the National Advisory Committee on NAFTA Environmental Affairs for the U.S. Environmental Protection Agency. She also is a geologist and an art historian.

CARLOS A. RINCÓN works on behalf of the Environmental Defense Fund to improve air quality along the U.S.-Mexico border's international air quality management district in Cd. Juárez, El Paso, Texas, and Sunland Park, New Mexico. Rincón possesses twenty years of professional experience throughout the U.S.-Mexico border region as a businessman, academic and government official. He received his Ph.D. in Water Resources from the Instituto Tecnologico y de Estudios Superiores de Monterrey.

ROBERTO A. SÁNCHEZ-RODRÍGUEZ is the former program manager for transboundary issues at the Commission for Environmental Cooperation of NAFTA. He has worked for many years on border environmental and urban issues and chaired a research team preparing a book-length study of voluntary organizations and agreements on both borders. Recently he joined the faculty at the University of California, Santa Cruz.

CHRISTINE L. SHAVER is currently the chief of the Air Resources Division, National Park Service. Formerly a senior attorney with the Environmental Defense Fund, Shaver also worked for the National Park Service as Chief of Policy, Planning, and Permit Review Branch, Air Quality Division; assistant regional counsel, U.S. Environmental Protection Agency; and environmental protection specialist, Office of Transportation and Land Use Policy, U.S. Environmental Protection Agency.

KONRAD VON MOLTKE works on international environmental relations. He is a senior fellow at World Wildlife Fund in Washington, D.C., and adjunct professor of environmental studies and senior fellow of the Institute on International Environmental Governance at Dartmouth College. He is editor of International Environmental Affairs, a journal for research and policy.

PAUL R. WEST, Ph.D., FCIC, has been director of the School of Environmental Studies since 1987 at the University of Victoria, where he is jointly appointed in Chemistry and Environmental Studies. He is a charter member of the British Columbia Round Table on the Environment and the Economy (1990–94), a member of the Standards Council of Canada Advisory Committee on Environmental management Systems (ISO 14000), and past chair of the Board (1995–96) of the Canadian Council for Human Resources in the Environmental Industry.

JOHN D. WIRTH is Gildred Professor of Latin American Studies at Stanford University and is president of the North American Institute, a trinational public affairs nongovernmental organization based in Santa Fe. He serves on the Joint Public Advisory Committee for the Commission for Environmental Cooperation of NAFTA, and is contributing editor, with Robert L. Earle, of *Identities in North America: The Search for Community.* Wirth is at work on an environmental history entitled *Smelter Smoke in North America: Transborder Air Pollution and the Regional Commons.*

Additionally, several of these authors, including Peter Emerson, Tony Hodge, Mary Kelly, Richard Kiy, Ann Pizzorusso, and Stephen Mumme, are contributors to forums and publications of the North American Institute.

INDEX

acid rain, 5, 249, 264–65
Acid Rain Accord, 14, 16. *See also* U.S.-Canada Air Quality Agreement
Agenda 21, 26
agriculture, 40, 79, 221, 252
Agriculture Land Reserve (ALR), 103
air, 6; air pollution, *see* pollution: of air; air quality, 4, 56, 65, 66, 82, 126–27, 129, 133, 136–37, 139, 145–46, 148, 150–51, 166, 189, 192, 195–97, 199, 201; air quality management, 10, 37, 65, 137, 181–82, 184, 216; Mexico, 24, 130; monitoring, 127, 129, 136–37, 139, 145, 148
Air Force 408th Strategic Reconnaissance Wing, 222
Alaska, 7, 249, 262
Alberta, 269
Albuquerque, N.M., 13
Alert Citizens for Environmental Safety (ACES), 23, 229
Alley, Jamie, 6, 20, 53, 96, 251
Alliance of Border Communities, 97
Allied Signal, 261
Altamirano Pérez, René, 213, 215
Amazon basin, 255–56
American Lung Association, 128
Amistad Lake, 221, 223, 225–26;

Amistad National Recreation Area, 227
anencephaly, 23, 208
Angel Yunes, Miguel, 228
Angulo, Carlos F., 125, 216
Argentina, 255
Arizona, 15
Arizona-Sonora border, 14
Arizpe Sada, Rodolfo, 227
Army Corps of Engineers, 36
Assemblée des évêques du Québec (Assembly of Québec Bishops), 113
Association of Maquiladoras, 216
Association québécoise de biologistes, 113
Atkeson, Tim, 15
Audubon Society, 98

Baja California, 7, 15, 235, 237, 248, 264
Bannister, Geoffrey J., 164
Barron, Elaine, 140
Basel Convention on the Transboundary Movement of Hazardous Waste, 244
Bath, Richard, 166
batteries, 20
Beagle Channel, 256
Beaufort Sea, 7, 262
Belcher, Madge, 229

Bentsen, Lloyd, 222–23
Big Bend National Park, 5, 16, 189–91, 193–96, 199, 201, 259
big business owners, national federation, Mexico (COPARMEX), 173–74
bilateral efforts and agreements, 14, 33–34, 45, 126, 136–38, 166, 172, 182, 190, 196–97, 200, 223, 227, 229–30, 234, 238, 240, 258, 262, 264–65, 269
Binational Air Working Group, 139, 142–44
biodiversity, 82, 189–90; Convention on Biological Diversity, 1993, 40
birds: migratory, 7, 39, 86, 111, 198; sanctuaries, 6, 248; waterfowl, 39
Black Gap National wildlife Refuge, 190
Blackman, Allen, 164
Blanco, Herminio, 194
Boeing, 101
Bonilla, Henry, 228
Bonoir, David, 249
Border Environmental Protest Rally, 226
Border Environment Cooperation Commission (BECC), 5, 9, 19, 22, 24, 36–37, 137, 268
Border Zone Task Force (BZTF), 213–14, 217
Borja, Diana, 194
Bouchard, Lucien, 111
Boundary Waters Treaty, 1909, 90
Bourassa, Robert, 109
Brackettville, Tex., 228
Brazil, 258
brickmakers and brickmaking, 165–68, 173–80, 183, 269; brick kilns, 166–68, 170, 175–77, 179–80, 182–83; fuel, 166–67, 169–70, 174, 179–80
British Columbia, 6, 53, 55, 84–85, 90, 93–97, 103, 262; Ministry of Environment, 55–56, 65, 95; population, 81
British Columbia–Washington Environmental Cooperation Council (ECC), 20, 53, 58–62, 64–67, 69–70, 92, 95, 96, 100–101, 104, 251, 266–67

British Nuclear Fuels (BNF), 263
Bureau of Reclamation, 36
Bush, George, 17, 18, 171, 223
businesses, 25, 44, 142–43, 146–49, 170, 172, 175–76, 178, 180, 182, 190, 192, 208, 212, 214–15, 217–19, 234–36, 241, 252, 253, 255–56, 258–59, 261, 268, 270; corporate responsibility, 209; environmental partnering, 217–18; minimize the risk of future environmental liability, 210
Bustamante, Albert, 222

Caldwell, Lynton Keith, 9–10, 22, 27n 5
California, 263
camel corps, 221
Canada, 5, 16, 114, 116, 244, 250, 257, 259, 261, 268; provinces, 4, 18
Canada-U.S. border, 9, 13, 35, 262, 268
Canada-U.S. Free Trade Agreement (FTA), 16
Canadian Arctic Resources Committee, 114
Canadian Council for International Business, 211
Canadian Council of Ministers of the Environment (CCME), 58, 71n 6
Cananea, Sonora, 37
Cañon de Santa Elena area, 190, 199
Carabias, Julia, 195
Carbón I, 189–93, 196–201
Carbón II, 189–201, 205n 46, 259, 265
Carbón III, proposed, 199
Central America, 16
Chambers of commerce, national federation, Mexico (CONCANACO), 173
Chamizal dispute, 12, 234
chemical spills, 225, 242, 245, 254
Chemical Waste Management (Chem Waste), 235, 239–42; commercial hazardous waste landfill, proposed, 224–27, 229, 238; hazardous waste disposal facility, RIMSA, 236, 259; Tratamientos Industriales de

Tijuana, S.A. (TITISA), 23, 234–43, 245, 248, 260
Chihuahua: desert, 189, 199; Ecological Law of the State of Chihuahua, 1991, 131
Childress, Marion, 225
Chile, 29n 21, 254, 257
Chinati Mountains, 190
Chirinos, Patricio, 236–37
Chisos Mountains, 189
chlorofluorocarbons (CFCs), 6, 215
Christopher, Warren, 141
chronology, 275–85
Citizens Against Nuclear Trash (CAN'T), 263
Citizens Against Radioactive Environment (CARE), 229
Citizens' Environmental Advisory Committee to the El Paso City Council, 170
Ciudad Acuña, Coahuila, 5, 194, 221, 226–27, 230
Ciudad Juárez, Chihuahua, 10, 11, 37, 125–29, 139–41, 147–48, 165–67, 169–70, 172, 174–76, 179–80, 182, 215–16, 218, 234, 250, 252, 265, 268–69; Brickmakers' Project, 141, 164–65, 169–70, 172, 179–82, 184, 260
Clean Air Act, 1990, 16, 132, 133, 135, 137, 147, 199, 249, 265
Clean Cities Program, 141
clean technologies, 176, 178
cleanup, 225
Cliche, David, 5, 108, 111, 120n 5, 253
Clinton, Bill, 18, 140, 258, 263
Coahuila, 191–92, 226, 228; Ecological Department, 227; State Legislature, 223, 227
coal mines, 16, 193–95
Coatzacoalcos, Veracruz, 259
Colorado River, 12, 14, 251
Colosio, Luís Donaldo, 266
Columbia Pacific Foundation, 97
Columbia River, 35, 54, 56, 58, 60, 66, 79

Comisión Federal de Electricidad (CFE), 191–92, 196
Comité de Defensa Popular (CDP), 169
Commission for Environmental Cooperation (CEC), 4, 8, 10, 25, 49n 2, 68, 70, 137, 191, 196–200, 211–12, 244, 248–49, 252, 258, 265, 267–68; citizen submissions, 197; secretariat reports, 197–98
Committee for the Defense of the Health of Ilo, Peru, 257
community, 99, 151, 174, 229–30, 239, 242, 245, 250, 258, 263; Alliance of Border Communities, 21; community right to know, 239, 245, 249–50, 253, 259, 268; North American community, 5, 8–9, 21, 27n 1, 219, 247, 251, 267, 270
Confederación de Trabajadores Mexicanos (CTM), 168
Confederation of Mexican Chambers of Industry (CONCAMIN), 211
conservation, 14, 40, 214, 243, 255; goals, 14; of energy, 214, 217, 219; of hazardous waste, 243; recycling, 214, 217, 219, 243; of solder, 214–15; of styrofoam, 214; waste reduction, 214, 217
continental market, 12
Coon Come, Matthew, 111
Coopératives d'économie familiale, 113
copper mines and smelters, 15, 262; emissions, 15, 37, 166; pollution, 23
Corpus Dam, 255
courts, 257; California Superior Court, 20; legal standing, 20, 250; review of environmental regulations, 93
Cuyahoga River: fire, 83

dams, 75, 269
DDT, 16, 252, 265
Deep Underground High Explosive Test (DUGHEST), 224; Mexican opposition to, 223; U.S. opposition to, 222
de Hoyos, Emilio, 227

INDEX

295

de la Madrid, Miguel, 14
de la Vega, Guadalupe, 170
Del Rio, Tex., 6, 23, 221, 223–25, 227–29, 231, 248; City Council, 222, 225
Denver, Colo., 13
Department of Defense, 222
Department of the Interior, 5, 7
deregulation: U.S. electrical transmission lines, 5
Díaz, Porfirio, 14
Dos Repúblicas, 194–95
Douglas, Ariz., 37
Dryden, Tex., 223, 224, 229–30, 238, 248
Dumas, Lucie, 108
dumping, 208, 216

Eagle Pass, Tex., 194, 225, 228
ecological perspective, 9, 270n 2
economy: growth, 12, 77–78, 83, 102–103, 151, 183, 199–200, 241–42, 249, 253, 270; informal sector, 164–65; management, 81
ECO-TEC, 176, 179
educating the public about pollution, 150, 181, 183, 192, 201, 208, 215–16, 224, 227, 229–30, 239–40, 253
Edwards Aquifer, 224, 226
1889 Convention, 11
Eisenhower, Dwight D., 223
electricity market, international, 119, 196; infrastructure development, 196
electric utilities, 16, 191
Electronics Industries Association, 216–17
electrostatic precipitators (ESPs), 191
El Paso, Tex., 10, 11, 37, 125–29, 132, 134, 139–40, 148, 165–67, 171, 178, 216, 234, 252, 265, 268
El Paso Natural Gas, 141, 173–74, 176, 179–80, 182
Emerson, Peter M., 10, 125, 259
emission control equipment, 191, 193, 195, 198, 201; ADVACATE, 195

emissions, 129, 131–36, 141, 145–46, 164, 166, 175–77, 189, 191–93, 197, 219
emissions trading programs, 16, 136, 138, 140, 147–50
endangered species, 40, 64, 194, 262
England, 244
environmental activism, 26
Environmental Awareness Protest March and Rally, 228
Environmental Defense Fund (EDF), 37, 139–41, 148, 197, 219
environmental impact statement (EIS), 195, 197, 237, 239, 241, 243, 254–55, 265, 268
environmental justice, 224, 229, 254, 256, 263
Environmental Law. See General Law for Ecological Equilibrium and Protection, Mexico, 1988
environmental management, 25, 34, 93, 102, 247, 250, 260, 261, 267, 269–70; transborder, 7, 12, 32–34, 70, 181–84, 210–12, 215–19
environmental nongovernmental organizations (ENGOs), 5–8, 12, 15, 18–19, 21–25, 30n 31, 44, 69, 70, 98, 112, 117, 126, 142, 172, 181, 183, 191, 193, 195, 198, 200–201, 229, 234, 237–40, 241, 245, 248–50, 252, 254–55, 258–61, 265, 269–70
environmental protection, 26, 219, 247, 253
Environmental Protection Agency (EPA), 5, 13, 19, 36–37, 63, 97, 133, 135, 139–40, 147–48, 166, 172, 174, 192–95, 197–98, 209, 213, 215, 217, 226–28, 230, 235–36, 243–44, 249, 260, 263; Green Lights program, 215; 33/50 program, 215, 260
environmental stress, 7
environment as a political issue, 7, 171, 181, 198, 209
Environment Canada, 12, 24, 95, 97
erosion, 82

INDEX
296

Estrada Orihuela, Sergio, 6, 234–35, 237, 240
Eugene, Oregon, 11
European Economic Community (EEC), 15
eutrophication, 77, 83
Exportadora de Sal, 7

Federal Attorney General's Office for Environmental Protection, Mexico (PROFEPA), 177
federal governments, 10, 88, 92–95, 126, 130–31, 133, 142, 147, 151, 171, 181–82, 184, 191, 194, 200, 216, 223, 226–27, 230, 237, 247, 250, 256, 260, 267–68
Federal Ministry of Commerce and Industry, Mexico (SECOFI), 170, 172, 194
Federal workers' housing agency, Mexico (INFONAVIT), 173–74, 178, 180
Fédération québécoise de la faune (FQF) (Quebec Wildlife Federation), 113
Finland, 244
fires, 225
Fish and Wildlife Service, 36
fisheries, 6, 64, 77, 79, 86, 89, 90, 256, 262, 266; beluga whale, 7; bowhead whale, 7, 262; California grey whale, 7, 240; Kemp's Ridley (sea turtle), 6; Loggerhead (sea turtle), 6; marine breeding ground, 6, 240; salmon, 6, 54, 70, 74, 88, 90, 95; sea lamprey, 89; shrimp, 6, 86; trout, 89; tuna, 6; tuna-dolphin dispute, 6
Flynn, Richard M., 110
forests, 82, 103, 254, 259; fires, 40
Fork Stockton, Tex., 225
Franco Barreno, René, 170
Franklin Mountains, 127
Fraser River, 78, 85–86, 90, 94; gold rush, 79
Fraser Valley, B.C., 21, 54, 97
Free Trade Agreement of the Americas, proposed, 258

Frei, Eduardo, 254
Frente Nacional de Organizaciones Ciudadanas (FNOC), 168
Freon, 6
Friedrich Ebert Foundation, 248
Friends of the Earth, 98
Fundación Ecológica Mexicana, 227

Gardner, Booth, 59
Garrison, John, 249
Gas Research Institute, 173, 179
General Agreement on Tariffs and Trade (GATT), 197
General Law for Ecological Equilibrium and Protection, Mexico, 1988: referred to in the text as the "Environmental Law," 130–31, 169, 171, 212, 235, 242, 249, 259, 268; revision of, 271n 5, 274n 34
geographical information survey, 10
Geographical Information System atlas, 242
Georgia Basin, 6, 56, 58, 60, 63, 72, 78, 81, 84, 87–88, 92, 94–103; identity, 101
Georgia Strait, 79
Georgia Strait Alliance, 69, 98
Gilbreath, Jan, 197
Glacier Bay, 249, 262
global commons, 14
global ecosystem, 7
Goldwire, Lloyd, 229
governance, 151; charts for the Great Lakes basin ecosystem, 88, 89; failure, 137–38
government induced nongovernmental organizations (GINGOs), 24
government organized nongovernmental organizations (GONGOs), 24
government regulation, 92–93, 126, 132, 137, 171, 174–77, 182–84, 194, 198, 200, 208–209, 212–14, 217–19, 222, 226, 235, 237, 240–41, 243, 254–57, 260; command and control techniques, 25, 93, 164; comparison of U.S. and Canadian

government regulation (*cont.*)
  legislation and regulation, 93; comparison of U.S. and Mexican legislation and regulation, 130–36; regulatory relief for businesses, 26, 84
Gramm, Phil, 222–23
grass roots activism, 22, 23, 110, 181–82, 200–201, 223, 230, 234, 240, 247, 248, 253–56, 261, 269
Great Lakes, 4, 5, 13, 16, 35, 72, 75, 83, 86–90, 94, 96–103, 108, 249, 251–52, 265, 267; Great Lakes Charter, 1985, 108; Great Lakes Fisheries Commission (GLFC), 8, 20, 88, 90, 99–101; Great Lakes–St. Lawrence Mayors Conference, 97; Great Lakes United (GLU), 8, 20, 98, 260; Great Lakes Water Quality Agreement (GLWQA), 1978, 14, 20, 35, 83, 91–92, 99, 102, 267; identity, 101; population, 81
Great Whale Project, 108–10, 115, 117, 248, 255; business support for, 112; Great Whale Forum, 112–13, 268; infrastructure construction, 110, 112; labor support for, 111; opposition to, 111–12; public review of, 116–18
Great Whale River, 110, 113
Greenpeace, 98, 119, 237, 244
Group of 100. *See* Grupo de los Cien
Grupo Acero del Norte (GAN), 192–93
Grupo de los Cien, 6, 23, 250, 269
Grupo Peñoles, 174, 178
Guadalajara, Jalisco, 250
Guanajuato, 198, 242
Guerro Negro, 240
Gulf Coast Waste Disposal Authority, 269
Gulf of Maine, 20, 256, 262, 264
Gulf of Mexico, 5, 226, 253, 260, 262; Business Council for Sustainable Development of the Gulf of Mexico, 256, 260
Guradi, Guido, 254

Gurria, Angel, 141
Gutierrez, Alfredo, Jr., 6, 221, 263
Gutierrez, Antonio, 20

Hanford Nuclear Reservation, 54, 79
Harcourt, Mike, 59, 63
hazardous materials, 130–31, 209, 225, 229, 254, 262; agricultural chemical waste, 77, 84; biomedical waste, 56, 60; dioxins, 66, 85, 237; dumping, 208; furans, 66, 85, 237; heavy metals, 66, 85, 166–67; industrial hazardous waste, 6, 60, 77, 131, 208, 214–15, 217, 219, 235–37, 240, 242, 261; nuclear waste, low-level, 6, 23, 225; radioactive waste, low-level, 6, 229; toxic waste, 23, 64, 78, 82–86, 91, 96, 102, 208, 223–24, 226; tracking system, 243; transportation of, 6, 56, 236, 243, 245; waste incineration, 21; waste sites, 20, 23, 84
hazardous materials disposal: contractors, 208; by export, 244; importation of, 244
hazardous waste sites and facilities, 224, 226, 228–30, 234, 238, 241–42, 245, 248, 256, 259, 268–69; Dryden low-level nuclear waste landfill, 225, 238, 248; illegal, 235, 242; Spofford low-level nuclear waste landfill, 20, 224, 225, 248, 263; Swan Hill incinerator, 269
Haztraks, 243
Hidalgo, 242
Hidrovía barge canal, 255, 272n 15
Hobby, Bill, 194
Hodge, R. Anthony, 5, 9, 72, 267
House Committee on Energy and Commerce, 193
housing, 11, 215; self-help, 214
Houston, Tex., 16, 269
Howekamp, David, 139
Hudson Bay, 114, 117
Hudspeth County, Tex., 224, 226, 229
Hueco Mountains, 127

human health, 14, 19, 38, 78, 125, 150, 175, 178, 183, 193, 200, 209–10, 213, 214, 227–29, 238–39, 242, 245, 256, 266, 269; illness, 11
Hydro-Québec, 109–11, 113–17, 119, 262

Imataca Natural Reserve, 256
Imperial Beach, Calif., 14
indigenous populations, 256, 262–63; Arctic Council, 263; Campo Indians, 263; Cree Nation of Canada, 5, 109–15, 117, 269; Cree Regional Authority, 115; Cree Regional Board of Health and Social Services, 115; Cree School Board, 115; First Nations, 70, 75, 97–98, 112, 114–15; Fort William Band, 97; Inuit, 113–15, 117, 262, 263; Inuit Circumpolar Council (ICC), 263; Inupiat, 262; James Bay Crees, 109; Kativik Health and Social Services Council, 115; Kativik Regional Government, 115; Kativik School Board, 115; land claims treaties, 97, 262; Maya, 263; Native Americans, 70; Pigeon Band, 97; trapping and harvesting rights, 115; water resources, 97
industrial salt, 7, 240, 248, 253, 268
industries, 75, 77, 79, 86, 129, 147–49, 166, 173, 175, 189, 195, 209, 218, 235–36, 241–42, 245, 259–62, 265, 268; dirty, 209; North, 12; South, 12
Instituto Nacional de Ecología (INE), 139, 235, 241–43
Instituto Tecnológico y de Estudios Superiores de Monterrey, 173–74, 176, 182
Integrated Environmental Plan for the U.S.-Mexican Border (IBEP), 171
integrated resource planning (IRP), 113–14, 116–17
International Association of Fisheries and Wildlife Agencies, 43
International Association of Great Lakes Research (IAGLR), 99, 101

International Boundary and Water Commission (IBWC), 13, 35–38, 223, 268
International Boundary Commission, 13
International Bridge, 227
International Finance Corporation (IFC), 193
International Framework Convention on the Reduction of Greenhouse Gas Emissions, 26
International Joint Commission (IJC), 7, 10, 11, 13, 28n 12, 35, 36, 57, 66, 84, 88, 91, 99–101, 104, 268; Air Quality Board, 66; diplomatic management of disputes, 36
International Waterways Commission, 13
Inuit and Circumpolar Studies Center, 113
Irrigation Congress, 13
ISO 14000 global standard, 25, 211–12, 217–19, 220n 3, 260, 268

James Bay, 114
James Bay 1, 114
James Bay 2, 116, 118. See also Great Whale Project
James Bay and Northern Québec Agreement, 1975, (JBNQA), 110, 114–16
Joint Public Advisory Committee of the CEC (JPAC), 8, 22
Jones, James, 195
Juan de Fuca, Straits of, 5, 54, 56, 90
Juárez Mountains, 127

Kelly, Mary, 5, 15, 189
Kennedy, Robert, 111
Kinney County, Tex., 226, 229
Kirton, John, 32, 41, 44
Kiy, Richard, 3, 6, 247
Kootenay River, 35

Lacondona rainforest, 23, 269
La Grande complex, 114
Laguna Madre, 6, 23, 264

Laguna Verde, 269
Lake Champlain, 108
Lake Erie, 7
Lake Memphrémagog, 108
Lake Michigan, 251
Lake Ontario, 84
Lake Pátzcuaro, 255
Lake Roosevelt, 54, 56, 60
Lake Texcoco, 269
La Paz Agreement, 1983, 5, 14–17, 36, 126, 136–37, 139, 196, 199–200, 202n 12, 226–28, 236, 238, 243; emergency response agreement (Annex II), 15, 266; international air quality management district (IAQMD) (proposed Annex VI), 140, 141, 153–57; Joint Advisory Committee on Air Quality Improvement for the El Paso–Ciudad Juárez–Doña Ana County Air Quality Management Basin (IAQMB) (Appendix I to Annex V), 141–46, 148–51, 158–61; smelter annex agreement (Annex IV), 15, 37; transboundary movement of hazardous waste agreement (Annex III), 15, 236, 241, 243; urban air pollution agreement (Annex V), 15, 37, 38, 48, 136–37
Laughlin Air Force Base, 221, 222
Laurier, Wilfred, 14
lead: and zinc smelters, 66; dumping slag, 20
least-cost analysis, 116–17
limestone, 222, 226, 229
linkage of trade and environment issues, 4, 6, 15, 17–18, 201, 209, 219, 256–58
locally undesirable land use projects (LULUs), 224, 225, 230, 231n 4
Lopez Mateos, Adolfo, 223
Los Alamos National Laboratory, 141, 173–74, 179–80, 182
Los Angeles, Calif., 7
Louisiana, 262–63
Lowry, Mike, 63

Lynch, Linda, 229

Maderas del Carmen area, 190, 193, 196, 199, 201
Maderas del Carmen Mountains, 189
Manitoba Hydro, 114
Manufacturing industries, national federation, Mexico (CANACINTRA), 173
maquiladoras, 4, 11, 17, 23, 25, 129, 148, 208–209, 212–14, 219, 230, 235–37, 241, 243, 261, 268
maquilas. *See* maquiladoras
Marine Science Panel, 69, 84
markets, informal, 165, 175, 178, 181–84, 260
Massachusetts, 20
Matamoros, Tamaulipas, 219
Matamoros/Brownsville, 23; contingency planning, 266
McDonald's, 219
McKinsey, Lauren, 41
Mercosur, 26, 253, 257–58, 270
Metalclad, 241–42, 248, 259
Mexican Advisory Council for Sustainable Development, 8
Mexican American War, 1846–48, 234
Mexican Ecological Commission, 228
Mexican Embassy, 228
Mexican federal development bank (NAFIN), 172, 174
Mexican Federation of Private Health Associations and Community Development (FEMAP), 164, 167, 169–70, 172–74, 176–80, 182–84
Mexican labor unions, 165, 183–84
Mexican National Water Commission (CNA), 36
Mexican political organizations, 165, 168–69, 177, 184
Mexican Revolution, 14
Mexico, 16, 192, 199, 209, 219, 221, 223, 228–30, 240–45, 249, 251–52, 256, 259–60; Congress, 242, 249–50, 257, 259–61, 263, 265; Constitution, 130; economy, 217, 219,

236; government restrictions on the press, 24; population, 11
Mexico City, 7, 18, 242, 250, 253; airport, 23, 269; earthquake, 23
Midwest, 5, 16, 265
Mier Ayala, Ricardo, 227
migratory species, 21, 38–39, 240; breeding grounds, 7, 240; marine mammals, 111
military posts, 221
Mina, Nuevo León, 236
Mission Energy, 192–93
Mitsubishi, 7, 240
Molten Metal Technology, 259
Monterrey, Nuevo León, 17, 189, 242, 250
Montreal Protocol, 6, 215
Movimiento Ecologista de Acuña, 229
Mumme, Steven, 32
Munton, Don, 32, 41, 44
Murkowski, Frank, 249

Nacozari, Sonora, 37
National Ambient Air Quality Standards (NAAQS), 132
National Association of Manufacturers, 216
National Environmental Policy Act, 197
National Park Service (NPS), 40, 189–90, 192, 226–27, 262
natural gas, 180, 185, 196, 197, 256, 262; Argentine-Chilean Joint Commission, 256; infrastructure investment, 180, 254
Nature Conservancy, 98
networking and information: BECCNet, 8, 22; CECNet, 8; corporate communications, internal, 213, 217; cross-border networks, 12, 227, 229, 240–41; faxes, 21; Internet, 21, 25, 65, 265
neural brain defect. *See* anencephaly
New Brunswick, 20
New England, 5, 265
New Hampshire, 20
New Mexico: Air Quality Control Act, 135

new source performance standards (NSPs), 133
New York, 5, 11
New York Power Authority (NYPA), 109–10
Niagara Falls, 11, 77, 251
Niagara Reservation, 11
Niagara River, 35
NIMBY (not in my back yard), 6, 224, 234, 242, 245, 248, 263, 268–69
Nooksack River, 54, 56, 60
North American Agreement for Environmental Cooperation, 4
North American Clean Air Alliance for Zero-Emission Vehicles, 48
North American Conservation Conference, 14
North American Development Bank (NADBank), 5, 8, 19, 21, 36, 137, 249, 268; loans, 24
North American Free Trade Agreement (NAFTA), 3-4, 8, 12, 17–19, 68, 83, 94, 125, 137, 139, 171, 173, 181, 191–93, 196–97, 200, 209, 211, 216–17, 228, 230, 238, 241, 244, 245, 248–49, 253, 257–59, 261, 264, 266–67, 270; environmental effects of, 23, 192; fast-track, 29n 21; opposition to, 29n 18, 171, 172, 249; overarching structure, 24; NAFTA Highway, 264; side agreements, 7, 191, 193–94, 196–97, 199, 206n 57, 207n 60, 250, 257
North American Fund for Environmental Cooperation (NAFEC), 8
North American Toxic Release Inventory, 10
North American Water and Power Project, 251
Nova Scotia, 20
nuclear facilities, 255, 269; nuclear submarines, 54, 63; threat of accidents, 54, 63
Nuclear Regulatory Commission, 263
Nuevo León, 189, 191

Núñez, Francisco, 140, 176–77

Official Mexican Standards (Normas Oficiales Mexicanas) (NOMs), 131
oil, 254, 256–57, 262; exploration lease rights, 7; refineries, 54, 166; spills, 54, 64, 67, 251, 262; supertankers, 54
Ontario, 11, 16
Ontario Hydro, 114
Organ Mountains, 127
Orozco Deza, Miguel Angel, 139
Ottawa, Ont., 18

Pacific Northwest Economic Region (PNWER), 100
Pacific Rim, 79, 95, 103
Pacific Salmon Treaty, 1985, 70, 90, 97
Pan American Health Organization (PAHO), 38
Pantanal bioreserve, 255
paper mills, 66, 79
Paradise Reef pier at Cozumel, 248–49, 252, 269
Paraguay River, 255
Parizeau, Jacques, 117–18
parks: management, 40, 190
Partido Acción Nacional (PAN), 169, 237, 250
Partido del Trabajo (PT), 169
Partido Revolucionario Institucional (PRI), 168, 237, 249–50
Paso del Norte Air Management Region, 10, 125, 128, 129, 136–40, 142–43, 148, 152, 166, 216, 250, 265–67; population, 128, 166
Paso del Norte Air Quality Task Force, 126, 140, 143–44, 146, 148, 150–51, 153, 216
Pataki, George, 249
PCB Management Plan for North America, 244
PEMEX, 23, 175, 240, 260
People for Puget Sound, 69, 98
People's March, 226, 227
PetroPac, 214

Phelps Dodge, 37
Philips Electronics North America Corporation, 210, 213, 216–19, 261; Airpax de México, 219; employees, 216; newsletters for environmental coordinators, 213; Philips Consumer Electronics Company, 214, 216; Philips Lighting, 216; policy statement, 213; Summit Componente, 219
PIBBY (place in black back yards), 263
Pichardo, Ignacio, 195
pictographs, 221
Piedras Negras, Coahuila, 5, 194
Pizzorusso, Ann C., 25, 208, 259–60
Playas de Tijuana, Baja California Norte, 235, 237–38
Playas de Tijuana Housewives Association, 23, 237–38, 248
polluter pays, 19
pollution, 83, 228, 257, 264; of air, 10, 38, 125, 129–33, 135, 137, 141–42, 145–47, 150, 165–67, 170, 181, 189–91, 198, 225, 238, 249, 254, 261, 264, 266; benzene, 128; carbon monoxide, 125, 127–28, 132, 135, 166–67; inversions, 127, 166; of land, 208, 229; land-based sources for marine environments, 10; Mexico, 5, 23; motor vehicles, 129, 261; ozone, 125, 127, 132–34, 166, 249; particulate matter, 125, 127–28, 132, 135, 166–67, 191–92, 227, 264; remediation projects in Mexico, 5; sources, 128; sulphur dioxide, 14, 133, 166–67, 189, 191–93, 195, 201, 257, 265; of water, 35, 83, 208, 222, 223, 225, 227, 254, 262, 268
polyaromatic hydrocarbons (PAHs), 86
polychlorinated biphanols (PCBs), 237–38, 240, 243, 244, 259–60
Pomeranz, Jeff, 222
Porcupine Caribou, 7
power plants: carbón power plants, 5,

16, 189, 191–92; coal-fired power plants, 5, 15, 189, 191–92, 249, 259; hydroelectric power projects, 5, 11, 91, 108–10, 113–14, 118, 251, 269
private voluntary organizations, 23
Projecto Fronterizo de Educacíon Ambiental, 23
propane, 169–70, 174–80, 182–84; price increase and subsidies elimination, 175–80
public advisory committees, 5, 8
Public Citizen, 18
public meetings, 61, 91, 113, 117–18, 142–43, 200, 242, 245, 253, 257
public-private partnerships, 26, 165, 172, 181–82, 191, 210, 216, 260, 268, 270
Public Utilities Regulatory Policies Act (PURPA), 1978, 109
Puget Sound, 5, 54, 56, 60, 63, 72, 78, 81, 84–85, 87–88, 92, 94–103; identity, 101

quality of life, 7, 83
Québec, 5, 108, 109, 117, 248, 253, 262, 265, 268–69
Québec branch of the Confederation of National Trade Unions (CSN), 113
Quemetco, 20
Querétero, 242

Ramón, Jr., Jesús Maria, 227–28
Ranchers, 229
RAND Corporation, 21
Reagan, Ronald, 14–16
reasonably available control technology (RACT), 135
Red River, 265
Regional Environmental Commission, Chile (COREMA), 254–55
Responsible Care, 260
Richards, Ann, 194, 224, 228
Rieff, Susan, 227
Rincón, Carlos A., 125

Rio Bravo. *See* Rio Grande
Rio Conference, 1992, 15
Rio Grande, 6, 12–14, 126, 189–90, 221, 225–27, 229, 234, 251, 268; Wild and Scenic River designation, 190
Rio Grande Alliance, 38
Rocha de Diaz, Marta, 237
rodeos, 221
Ronfeldt, David, 21
Roosevelt, Theodore, 13–14
Rosenthal, Andres, 228
Ross, John C., 138
RSR Industries, 20

St. Croix River, 264
St. John's River, 35, 264
St. Lawrence River, 35, 75, 98, 108, 251
St. Mary's and Milk River, 35
Salinas de Gotari, Carlos, 17, 140, 172, 175, 190, 223
Samper Pizano, Ernesto, 257
Sánchez-Rodríguez, Roberto A., 8, 32, 253
Sanderson Citizens Against Toxins (SCAT), 229
Sanderson, Tex., 225
Sandia National Laboratory, 141
San Diego, Calif., 5, 7, 11, 14, 152, 238
San Diego Bight, 6
San Diego–Tijuana metropolitan region, 10; San Diego–Tijuana Emergency Planning and Response Committee, 266
San Felipe Creek, 222
San Felipe Springs, 221, 222, 226
San Ignacio Lagoon, 7, 248, 253, 268
San Juan Ixhuatepec, Mexico, 23
San Luis Potosí, Mexico, 241, 249, 259
scientists and scientific studies, 62–65, 69, 126, 129, 131–32, 139, 141, 183, 189, 196, 198, 201, 229, 240, 253, 265
Sea of Cortez, 253, 264
Searsport, Maine, 264
Seattle, Wash., 63

Secretariat of Ecology, Urban Development, and Housing, Mexico (SEDUE), 13, 209, 213–15, 227, 235–37, 241, 243
Secretariat of Environment, Natural Resources, and Fisheries, Mexico (SEMARNAP), 13, 36, 37, 40, 139, 142, 194–95, 209, 243
Secretariat of Social Development, Environmental Regulation and Enforcement, Mexico (SEDESOL), 172, 193–94, 209, 216, 243, 266; Green Flag Award, 216
Shaver, Christine L., 125
Sierra Blanca, Tex., 5, 23, 224, 226, 230–31, 238, 259
Sierra Club, 18, 98, 194–95, 244, 260; Legal Defense Fund, 244
Silva Reservoir, 198, 248, 269
Sindacato de Ladrilleros y Trabajadores de la Cal, 168–69
Skagit County, Wash., 21, 97
Skagit River, 35
social capital, 22
Soldaduras Omega, 215
Solidaridad (PRONASOL), 187n 27; Solidarity Enterprises (Empressas Solidaridad), 172–74, 176
Sonora, 15
Sony, 261
Sounds and Straits Coalition, 69, 98, 100
Southern California Edison, 193
Southern California Waste Management Authority, 269
Southern Peru Copper Corporation, 256, 273n 19
Southwest Center for Environmental Research and Policy (SCERP), 173–74, 178
sovereignty, 12, 252
Spalding, Mark, 21–22
Spence, Shirley, 226
Spofford, Tex., 223, 228, 230
Star Wars, 222
state governments, 10
state implementation plan (SIP), 134–35

state-provincial linkages, 44
state visits, 228, 258
subnational units of government, linkages, 41–43, 45, 67, 71n 4, 92–93, 95–97, 103, 126, 130–31, 133, 142–43, 146–47, 166, 172, 180–83, 191, 200, 210, 228, 238, 247, 249–51, 253, 256, 267–69; hazards monitoring, 47; provincial governments, 10, 93–94. *See also* Canadian Council of Ministers of the Environment (CCME)
Sunland Park, N.M., 125–26, 128–29
sustainable development, 8, 37, 85, 101, 102, 117, 190, 199, 241
Swanson, Roger, 41, 44

Taft, William Howard, 14
Tamaulipas, 264
Tatshenshini wilderness, 262
technical education, 174, 183–84, 212, 268
technology transfer, 174
Tepoztlán golf course, 240
Terrell County, Tex., 223–24, 226, 229
Texaco, 256, 272–73n 19
Texas, 5–6, 189, 192, 194, 199, 226, 229, 238, 248, 260; Railroad Commission of Texas, 195; Texas Air Control Board, 140; Texas Attorney General's Office, 197; Texas Clean Air Act, 134; Texas Department of Health, 225; Texas General Land Office, 194, 229; Texas Natural Resource Conservation Commission (TNRCC), 134, 139, 140, 147, 191, 194–95, 227–30; Texas Parks and Wildlife Department, 190; Texas Water Commission, 224–25, 227
Texas Center for Policy Studies, 6, 23, 229
Texas Low-Level Waste Disposal Authority's site, proposed, 224
Texcor low-level nuclear waste landfill, proposed, 224, 225–26, 228, 230
Thorup, Catherine, 21

Tijuana, Baja California Norte, 5, 7, 11, 14, 20, 152, 235–38, 250
Tijuana River, 14
timber, 254; harvesting, 6, 75, 249
Toronto, Ont., 265
Torres, Adrian, 238
trade, 7, 12, 75–80, 95, 103, 109, 196, 209, 254, 264
Trail Smelter, 29n 23, 53
transboundary environmental impact assessment, 10, 116–18
transboundary environmental interactions, 32, 55, 57, 66–68, 70, 92, 98–100, 120, 126, 136, 140, 147, 151, 174, 223, 227, 229–31, 238, 240–41, 244, 251–52; academic research, 99, 100, 174; of air, 65, 125, 129, 138–39, 147, 150–52, 181; funding, 68, 84, 99, 104, 137, 144, 166, 171–72, 174, 176, 180–81, 191, 195–96, 198, 200–201, 248; infrastructure construction, 137, 235, 241, 254; management, 45, 164–65; public support of, 166; suspension of, 70, 97; trilateral, 34; of water, 35
transboundary environmental issues, 3–4, 83, 103–104, 129, 136, 138, 169, 171, 181, 190–91, 193–94, 196, 199–200, 208, 222, 228–30, 234, 239, 243, 245, 249, 251–54, 257–58, 260–61, 264, 267–69
Transboundary Resources Inventory Program (TRIP), 10, 233n 17
TransCanada Highway, 264
transportation, 225, 261
trilateral efforts and arrangements, 45, 197, 200, 247, 249, 252–53, 270
Trillium Corporation, 254–55, 259

Union de producteurs agricoles (UPA), 113
Union québécoise pour la conservation de la nature (UQCN), 113
United Nations, 15; Working Group on Indigenous Peoples, 115

United States, 16, 228, 244, 251, 259–61, 268; Congress, 19, 217, 228, 230, 257; House of Representatives, 19; retirees living in Mexico, 238; U.S. State Department, 10, 18, 248, 266
Universidad Autónoma de Ciudad Juárez, 173
University of Arizona's Udall Center, 22
University of Texas at Austin (UTA), 197
University of Texas at El Paso (UTEP), 138–39, 173, 180, 182
University of Utah, 180
uranium mines and mills, 224–25
U.S. Agency for International Development, 21
U.S.-Canada Air Quality Agreement, 14, 48. *See also* Acid Rain Accord
U.S.-Canada-Mexico Migratory Bird Treaty, 14
U.S. Council for International Business (USCIB), 211, 216
U.S. Defense Nuclear Agency (DNA), 222–23
U.S. Good Neighbor Environmental Board, 8
U.S. Gulf intracoastal waterway, 6, 264
U.S.-Mexico border, 4, 19, 37, 125, 166, 184, 189, 193, 232n 12, 234, 238, 242, 254, 260, 263
U.S.-Mexico Border Environmental Cooperation Agreement, 1983. *See* La Paz Agreement
U.S.-Mexico Border Governors Conference, 238
U.S.-Mexico Border Health Commission, 38
U.S.-Mexico Border XXI Program, 6, 8, 19, 137
U.S.-Mexico Free Trade Agreement, 17
U.S.-Mexico Integrated Border Environmental Plan, 113–14, 116–17, 226
U.S.-Mexico Water Treaty, 1944, 14, 36, 226, 251, 268

Val Verde County, Tex., 226

Vancouver, B.C., 11, 63, 85, 95
Victoria, B.C., 5, 63, 85
Villarreal, Francisco, 176
visibility, 189, 192–93, 195, 259
Von Moltke, Konrad, 32

Washington, 6, 21, 53, 55, 84–85, 90, 96; Department of Ecology, 55–56, 65; population, 81
Washington, D.C., 18, 228, 230
waste disposal: hazardous waste management, 208, 225; liquid waste management, 63; sewage, 5, 7, 13, 14, 54, 63, 65, 77, 83–85, 95, 222, 253, 261–62; solid waste disposal systems, 11, 56, 60, 137, 263; wastewater treatment, 11
waste minimization, 213
water, 5, 23, 268; agricultural contamination of groundwater, 54, 252, 262; groundwater, 54, 56, 60, 194, 225, 227, 230, 234, 251–52, 263, 268–69; potable water, 11, 215; water allocation, 12, 35, 91; water quality, 4, 13, 35, 54, 56, 60, 63, 65, 66, 82, 84, 91, 251; water quality management, 36, 56, 60, 251; water supply in Del Rio, Tex., 221, 223; water supply in the Valley of Mexico, 24
Watson, Kirk P., 140
West, Paul R., 5, 9, 72, 267
Western Governors Association, 8
Western Wildlife Health Cooperative, 48
wetlands, 40, 64, 82, 100, 255, 262
Whatcom County, Wash., 21, 97
wildlife protection, 39
Windy Craggy project, 262
Wirth, John D., 3, 29n 23, 247
World Bank, 24, 193, 242
World Trade Organization (WTO), 6, 197, 257–58

YIMBY (yes in my back yard), 269
Yukon Territory, 262

Zedillo Ponce de Leon, Ernesto, 130, 140, 196